Industrial Strength Software

Effective Management Using Measurement

IEEE Computer Society Press
Mohamed E. Fayad
Editor-in-Chief, Practices for Computer Science and Engineering

Industrial Strength Software

Effective Management Using Measurement

Lawrence H. Putnam
Ware Myers

IEEE Computer Society Press
Los Alamitos, California

Washington • Brussels • Tokyo

Library of Congress Cataloging-in-Publication Data

Putnam, Lawrence H.
  Industrial strength software: effective management using measurement / Lawrence H. Putnam, Ware Myers.
      p.   cm.
  Includes bibliographical references and index.
  ISBN 0-8186-7532-2
  1.   Computer software—Development—Management.
I. Myers, Ware.   II. Title.
QA76.76.D47P867   1997
005.1—dc21   96-37019
                                                        CIP

IEEE Computer Society Press
10662 Los Vaqueros Circle
P.O. Box 3014
Los Alamitos, CA 90720-1314

IEEE Computer Society Press Order Number BP07532
Library of Congress Number 96-37019
ISBN 0-8186-7532-2

Additional copies may be ordered from:

IEEE Computer Society Press	IEEE Service Center	IEEE Computer Society	IEEE Computer Society
Customer Service Center	445 Hoes Lane	13, Avenue de l'Aquilon	Ooshima Building
10662 Los Vaqueros Circle	P.O. Box 1331	B-1200 Brussels	2-19-1 Minami-Aoyama
P.O. Box 3014	Piscataway, NJ 08855-1331	BELGIUM	Minato-ku, Tokyo 107
Los Alamitos, CA 90720-1314	Tel: +1-908-981-1393	Tel: +32-2-770-2198	JAPAN
Tel: +1-714-821-8380	Fax: +1-908-981-9667	Fax: +32-2-770-8505	Tel: +81-3-3408-3118
Fax: +1-714-821-4641	mis.custserv@computer.org	euro.ofc@computer.org	Fax: +81-3-3408-3553
Email: cs.books@computer.org			tokyo.ofc@computer.org

Editor-in-Chief: Mohamed Fayad
Publisher: Matt Loeb
Acquisitions Editor: Bill Sanders
Developmental Editor: Cheryl Smith
Production Editor: Lisa O'Conner
Advertising/Promotions: Tom Fink
Figures and tables by Douglas Putnam

Printed in the United States of America by Braun-Brumfield, Inc.

The Institute of Electrical and Electronics Engineers, Inc

Contents

Foreword

On a recent consulting assignment at a major high tech company, I stopped by the cubical of a young software engineer. I noticed that his cork board was covered with a large hand drawn Dilbert comic strip. Nodding my head in the direction of the cork board, I asked, "Is that a copy or an original?"

He smiled and looked sheepishly at the floor. "I drew it myself," he said without meeting my eyes, "And one of your books gave me the idea."

"Really," I said, not sure whether to be flattered or mortified. I got up to take a closer look.

It was a reasonably good forgery, a strip with three cells. In the first, the Dilbert character is in the boss's office, with another more senior looking manager who is standing behind the boss. The boss says, "I'm briefing Mr. Big on our software project. How's it coming?"

In the second cell, the Dilbert character states, "Well, we've run into a few serious problems. We'll never get done by the deadline. We all agreed that the deadline was too short anyway." The boss looks concerned. The senior manager looks bored.

In the third cell the boss responds. "OK, so we've run into a little trouble, but nothing that a few long evenings won't cure." Dilbert looks sick. The senior manager is leaving the room. He looks back at Dilbert and his boss and says, "Oh, I forgot to tell you. We'll need the project done two weeks early. I'm sure you'll be able to do it."

I turned back to the young engineer. "Is that the way things are around here?"

He just smiled.

Many of us who have spent decades managing software projects (and later writing about our experiences) have lamented the lack of precision with which we plan, track, and control these projects. Software, it seems, is almost always late, is frequently over budget, and is all too often unreliable. We shrug our shoulders and shake our heads as we move to the next project, hoping that we'll have the wisdom to make things better.

In their book, *Industrial Strength Software*, Larry Putnam and Ware Myers provide the wisdom that can help you to make things better.

Their approach combines substantial doses of common sense; a pragmatic insistence that you must measure in order to manage effectively; the realization that software engineering, like most other things in real life, has rhythms and patterns that can be understood if properly analyzed, and an optimism (backed up by copious industry data) that things can get better.

Throughout this book, the authors propose a simple approach to software project management: (1) understand key concepts for each management activity; (2) find metrics that quantify these concepts; (3) measure; (4) find a pattern based on the measurements; and (5) respond using the pattern as a guide. This approach applies regardless of whether you're managing a project, working to improve product reliability, or attempting to proceed to a higher level of software process maturity.

I think Dilbert would like this book, even if it did result in better management and less grist for the comic strip. He'd enjoy the anecdotes, the concise presentation, the pragmatic approach. He'd revel in a philosophy that encourages a realistic view of how to manage quantitatively. He'd even enjoy the modular presentation and at least some of the graphs. Unlike Dilbert, Larry Putnam and Ware Myers don't use comic strip artwork to tell their story, but it's a tale that's well worth reading.

Roger S. Pressman
December 1996

Preface

The book you now hold in your hands is aimed at all the participants in software development: executives, managers, supervisors, technologists, analysts, coders, testers, and documentation people. You have a common interest in the effective management of projects, the attainment of reliable products, and the continuing improvement of the software process itself. These are the three aspects of software management with which this book is principally concerned.

Some of you, such as general executives, may have no particular background in software. You may be vice presidents with responsibility for several functional areas, one of which is software; division general managers directing all the functions within a profit center; and chief executive officers. You have risen to a level where you direct functions in addition to the one in which you had your own early experience. You did not yourself come up through the software function, but you now have it within your jurisdiction.

Others of you are presently grappling with the immediate problems of managing software development. Still others may not yet have had the experience of managing a complex activity such as software development. You may have little knowledge of managing a project over time, reducing defects, and investing in process improvement.

With a grasp of the three interlocked aspects of software development, we believe a general executive without professional software experience can better oversee this function. Obviously, in the limited time a busy executive life allows, you cannot learn the technology of this field in any detail. It is possible, however, to sort out the knowledge you need to operate effectively at your level. In this book we help you do that.

With this same three-way grasp, directors and managers who do have software experience can better manage the development process. At the working level people can see how they fit in the bigger picture and be better able to manage effectively when opportunity knocks.

Now, how does one manage any complex activity? In essence, you pick attributes to measure, you measure them, you keep accessible records of the measurements, you use the records to estimate attributes of new projects, you control against the estimate, and you make course corrections when actuals vary from estimates.

Unfortunately, software development does not progress in accordance with the rather simple rules that govern most functions. That is why software projects run beyond delivery dates by many months, overrun budgets not just by a small percent, but often by a large factor, and are even canceled about one-quarter of the time. That is why many software products are delivered with hundreds of defects and mean time to failure measured in a few hours. That is why as many as 75 percent of software organizations are mired in the chaotic state characterized by the Software Engineering Institute's Capability Maturity Level I as "the lack of a managed, defined, planned, and disciplined process for developing software."

To overcome this chaos, executives, managers, and software people need to concern themselves with three aspects of software development. One is the progress of individual projects. Are you completing them on time and within budget? The second is reliability. Is the resulting product sufficiently defect-free to be able to run long enough between failures to serve the needs of its application?

The third is the long-run improvement of the development process itself. Is your organization's ability to develop software improving year by year? Are you matching the progress being made by your competitors? As software becomes a greater part of the operating cost of enterprises, the ability to improve your process becomes a competitive advantage. Worse, the inability to keep up leads to the corporate graveyard.

You can reduce this third phase to the same measurement framework as the first and second phases.* You measure the efficiency of your present process, you invest some money to improve it, and you remeasure its efficiency. If efficiency has increased, great. Invest some more money—with confidence.

How to get a good measure of the overall productivity of the development process has not been obvious. Many of the measures that have been used in the past have proved to be grossly inaccurate. A good metric lets everyone see if they are doing the right things and doing them well.

This book seeks to convey an understanding of the three phases of software development. In our earlier book, *Measures For Excellence: Reliable Software on Time, within Budget* (Prentice Hall, 1992), we provided more detailed information on how to apply these methods.

---

* We are using the word "phase" here, not in the sense of one in a series of stages (which is one of its dictionary meanings), but in the sense of one of the major interlocked, concurrent aspects of software development (which is based on its primary dictionary meaning).

One of us (Putnam) has been researching measurement, estimating, project control, reliability, and investment management since the mid-1970s. Out of this work he evolved the Software Life-Cycle Management system, SLIM, and related software tools, Size Planner, SLIM Control, PADS (Productivity Analysis Database System), and Programmer's Personal Planner. In the course of this work he has accumulated a database of 3885 projects that provide the statistical underpinning for three-phase software management.

Putnam initially analyzed software production from the vantage point of Special Assistant to the Commanding General of the Army Computer Systems Command. From the beginning he viewed software problems from a top-management position. This Command, then numbering about 1700 people, developed the software that operates the logistic, personnel, financial, force-accounting, and facilities-engineering functions at worldwide Army installations.

A graduate of the US Military Academy at West Point with an MS in physics from the US Naval Post Graduate School, Monterey, California, Putnam had leadership experience with troops early in his career. His knowledge of the theory of management was enhanced at the Command and General Staff College. His understanding of the practical problems of top-level management was extended by four years in the Office of the Director of Management Information Systems and Assistant Secretary of the Army at Headquarters, Department of the Army.

As president of Quantitative Software Management, Inc., McLean, Virginia, since 1978, he faces the same variety of problems that any independent business entity does, including the difficulties inherent in developing successive generations of the software products that implement three-phase software management. In particular, "I never have enough time to get everything done I would like to do," he says. His personal consulting relationships with people in major corporations, not only in the United States, but also in Europe and Asia, have given him a keen appreciation of the constraints under which they must develop software.

Myers first developed his understanding of the problems of making everything work together as production control manager of a rubber molding plant. A far cry from software—although the rubber was not exactly "hard" ware, the control of production did involve time and cost measurement, job estimating, production and material planning, tool planning, and tracking of production against plan. Years later, as a contributing editor of *Computer* magazine, when he first heard Putnam outline his ideas, he felt that Putnam had a significant concept. He brought it to a wider audience in a December 1978 article entitled "A Statistical Approach to Scheduling Software Development." Since then he has written many other articles on software management.

Myers is an engineering graduate of Case Institute of Technology, Cleveland, Ohio. His master's degree in administration is from the University of Southern California. For many years he taught a course in Engineering Organization and Administration to seniors and graduate students at the University of California, Los Angeles.

Serving as volunteer treasurer of several nonprofit organizations, Myers found that other board members tended to focus on the delights of spending money for well-meaning purposes while the treasurer, always conscious of the limited funds coming in, had to be the restraining hand.

Serving as the restraining hand is, in a nutshell, also the task of this interlocked approach to software development: getting work done with limited resources and time, enhancing the quality of the product while doing so, and continuously improving the way the work gets done. Competition never sleeps.

Lawrence H. Putnam
Ware Myers
December 1996

Part I

Management View of Software Development

Human life is a mix of work and play. When we are playing, we do not pay strict heed to the passage of time, our effort, or the efficiency with which we are playing. The pleasure of playing is its own reward. When we are working, we must heed the passage of time, for work has a due date. We must heed the effort taken, because work implies a cost target. We must give regard to efficiency, because the product of our work has to compete with the product of someone else's work.

Late at night, hunched over his home computer, a programmer searches for the next blockbuster application program. No one asked him to invent the greatest program since the spreadsheet. No one gave him a budget, a schedule, or a reliability target. He is playing.

During the day, sitting before his office computer, a programmer is coding a module passed on to him from detailed design, one of several hundred in a big system. A supervisor assigned this task, along with a due date. The programmer is working.

It is the programmer—or more broadly, the software developer—at work that is our concern. Management is the art of planning work so that it can be accomplished within constraints of time, cost, and other resources at a level of efficiency that will be competitive in the marketplace. In Chapter 1 we pose the questions for which management needs answers, and we outline the path to those answers.

Managers need answers. They have little choice, because modern organizations must have software. That is the theme of Chapter 2. They need it, not only to make operations more efficient, but also because, more and more, software is a strategic component of their operations.

Planning software development is in some ways analogous to planning manufacturing, but it is more complex. It is a subject we begin in Chapter 3 and continue in Part II.

Chapter 1

The Management Questions

"I don't have a lot of time," the division manager told us. "In addition to software development, I oversee marketing, product development, manufacturing, accounting, and other functions. I am interested in getting answers to practical questions such as these."

He handed us a single sheet of paper.

"These are the management questions," he said. "It helps my people focus on what I need to know to run the division."

The sheet contained a dozen questions:

- ❑ How big is the project?
- ❑ How long will it take?
- ❑ How many people will it employ?
- ❑ How much will it cost?
- ❑ Will the product satisfy the reliability needs of its application?
- ❑ What are the risks of not meeting cost, schedule, or reliability goals?
- ❑ Are there tradeoffs? What are they?
- ❑ Should we be doing the project at all?
- ❑ Where do we stand as an organization? Are we competitive?
- ❑ What is our productivity? What is our quality level?
- ❑ What must we do to get better?
- ❑ Can we make it pay off?

"Yes, we have encountered questions like these before," we said. "People can't answer them off the tops of their heads. You must have organizational practices established that bring up the numbers that answer the questions. In software development, establishing these

practices requires first measurement, and then the use of the measures to manage the three fundamental phases of software development."

How can I pick the beef out of the hash?

As professional fields have become more intricate, management has become the art of abstracting from this complexity a few primary concepts and measurements. A manager needs these concepts to ensure that the function is operating effectively and to manage its intersection to other parts of the organization.

Standard accounting procedures are an example of this abstracting technique. The quantity of financial detail in an organization of any size is mind-boggling. Out of this detail, the accounting function fashions, at the top level, profit and loss statements and balance sheets; at the intermediate level, cost center reports; and at the working level, reports of costs that exceed standards.

You need not understand the agonies the accounting department goes through to collect all this data, winnow it through complicated computer procedures, and generate reports centered on the needs of each management level. You must know merely how to apply this focused information to your own responsibilities.

Material control and production control are other examples of control mechanisms that pull information on which management needs to act out of a mountain of detail.

These managerial control activities are analogous to real-time control of a physical process, such as a chemical reaction. Fundamentally a control system measures a process's key variables, compares them with desired set points, and feeds back corrective action to the process.

The engineers who design the system have to know a great deal about the process. For example, they have to sort from a host of possible measurements the few that are sufficient for adequate control. They have to develop the algorithms that use these measurements to compute feedback to be applied to the process. The resulting automatic control system could hardly be said to have a deep understanding of the technology it is controlling. It operates on a few fundamentals abstracted from all the details of the process.

Another example is total quality management. It has been much in the news the last few years. Motorola, Xerox, IBM Rochester, and General Motors Cadillac have won Baldrige awards. This program improves the quality not only of the product but of all the processes, direct and indirect, that enter into its production. Thus, it embraces all functions of the enterprise.

No one could hope to understand these many processes in detail. Yet executives are trained in about a week in what they need to know to initiate and carry on the total quality management program. Managers and specialists are the ones who have to understand the many details in their own fields. We need an analogous program in "total software management."

What *is* the beef?

In concept the core ideas underlying the management of any economic activity are straightforward. First, we need to understand quantitatively what is going on:

1. Sort out the *key concepts* underlying the activity.
2. Find *metrics* that quantify these concepts.
3. *Measure* a number of samples of the activity.
4. Find the *pattern* in these measurements.

We call this the *measurement level*. In software development, key measurements include size, number of staff, staff hours or effort, cost, schedule, process productivity, and defect rate. These are the "management numbers."

Then we use these measurements to control three interlocked phases: (1) manage the activity; (2) improve the product's reliability; and (3) invest in the improvement of the process.

In the activity or *project-management phase*, the basic ideas are to

1. Lay out a plan in terms of the estimated management numbers.
2. Measure the actual numbers as the project proceeds.
3. Compare the actual numbers against the planned numbers.
4. Take action to bring execution in line with the plan.

In the *product-reliability phase*, the basic ideas are to

1. Project the estimated defect rate.
2. Find the defects at approximately the rate projected, or investigate the deviance.
3. Correct defects.
4. Correct deficiencies in the process that led to the defects.

In the investment or *process-improvement phase*, the basic ideas are to

1. Find an objective measure of the efficiency with which the organization carries out its activity. We call this metric "process productivity."

2. Determine the process productivity of the organization in the present (or recently completed) period.
3. Invest in process improvement.
4. Remeasure process productivity.
5. If remeasured productivity has, in fact, increased, continue to invest along the same lines. If remeasured productivity has improved little or none, rethink, regroup, and try again. And look over your shoulder. Someone may be gaining on you.

The methodology underlying these steps—*measure, do, remeasure, correct*—has worked in many fields, but it has proved difficult to apply in the software field. Useful metrics were not immediately apparent. The data collected was often poorly defined and "noisy," that is, highly variable and imprecise. The process of software development did not readily yield its inherent pattern. In particular, a good measure of process productivity did not emerge at first, and that measure underlies the steps in project management and process investment and affects product reliability.

Despite these problems, and as you will see later, there is a workable pattern. There is a sound measure of process productivity. Of course, both are quite complex, but you don't have to worry about that. We have worked out the measurement pattern. We have embodied it in software tools and they handle the complexity easily. That leaves you free to deal with the management questions.

Is there a software crisis?

At this point you may be inclined to say, and with considerable justice, "If you can get some good metrics, plan with them, and control against them, you can certainly manage any process. But we are now some 40 years into the software age and projects are running out of control all over the world. Didn't you guys ever hear of the software crisis?"

Well, yes, we have. We used to begin with some handwringing about the crisis. The idea was to get everyone in a mood to do something. By now, however, everyone knows about the software crisis. Most of you sense there is something—or a lot—wrong with the software work with which you are involved. And you know you want to do something about it.

Peter Freeman of Georgia Tech's College of Computing says the "software crisis" is dead! It is dead in the sense that the term "crisis" refers to a turning point and we have passed the turning point. "We have changed from feeling that software is some kind of totally unmanageable beast to believing that under the right conditions we can manage it just as we have learned to manage other problematical situations in our universe" [1].

Has anyone actually managed software successfully?

Our database of 3,885 completed software projects tells us that, yes, companies are managing software successfully. It also provides evidence that other companies are not.

For example, the ability to complete projects on schedule and within cost estimates is certainly one indicator of successful software development in the project-management phase. More than 600 of the projects reported to our database include both preliminary planning information and final schedule and cost data. About one third of these projects were completed within cost and schedule targets, one third were over target but within 125 percent of target, and one third exceeded 125 percent of target, as Figure 1-1 shows.

On the basis of this evidence we might say that around 200 projects were very successfully managed on these two indicators. Another 200 were managed fairly well in the sense that schedule slippage and cost overrun were well below crisis levels. The last 200, however, might be characterized as "out of control."

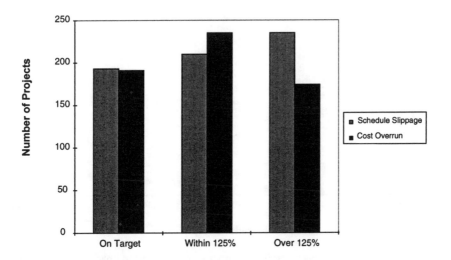

Project Schedule & Cost Overrun Statistics
from QSM Data Base

Figure 1-1. Among the organizations advanced enough to be reporting project data to our database, about one third are on target and about two thirds are not meeting their schedule-cost targets.

Our point is merely that some organizations seem to manage software development successfully. Others don't. We do not believe that the 200-200-200 pattern found in our database is representative of the entire universe of software organizations. In fact, assessments of software development organizations by the Software Engineering Institute of Carnegie Mellon University indicate that most of them operate without cost estimates and project plans.

Companies reporting data to us have been improving the reliability of their software, too. Figure 1-2 charts the decline in defects remaining at delivery since 1980 for a system normalized at 100,000 source lines of code. If progress continues at the same rate, the defects remaining will drop to 8.5 by the end of the century, or 0.085 per thousand SLOC. This figure is based on average business system data. Some companies developing real-time systems or other critical applications, in which the need for reliability is higher, are doing better. Companies not making measurements are probably doing much worse.

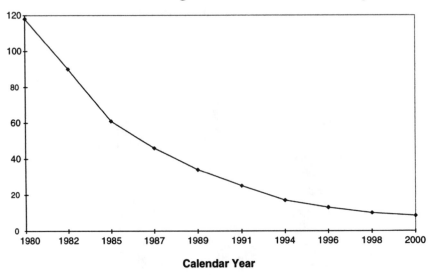

Figure 1-2. Companies with a software process disciplined to collect and report data have been making steady progress in reducing defects remaining at delivery since 1980.

In the investment or process-improvement phase, the rate of increase in process productivity for business software projects reported to our database between 1980 and 1990 averaged 11 percent per year. This rate of improvement means that schedules were being shortened by about 2.5 percent per year and effort (cost) was being reduced about 12 percent per year.

Figure 1-3 illustrates the highest process-productivity figure reported to us each year. The best organizations improved dramatically during this period.

For now, three points are clear: many companies—at least of those reporting data to us—are managing the cost and schedule aspects of software development projects effectively, but one third are not. Second, these reporting companies are improving the reliability of their products each year. Third, they are improving their software process each year. The point is our data is an existence proof that *improvement is possible.*

But it is not easy. If it were, more companies would be doing better. With an understanding of fundamentals, however, you can help make software development proceed more smoothly.

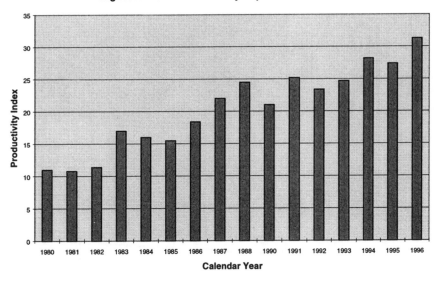

Highest Process Productivity Reported 1980 -1996

Figure 1-3. The highest process-productivity values reported to us over the past 16 years presented a rapidly moving target for less well managed competitors.

In January 1991, Motorola gathered a team of 27 officers and facilitators for a three-day training session on software. "I felt we could generate significant new business opportunities with improved software development performance," then CEO George Fisher said. "This group has taken leadership responsibility. I continue to be an active participant. I ask all Motorola executives and managers to seek the information and training needed to understand the software issues in their organization and contribute to the software solution. We need their commitment and personal involvement" [2].

References

[1] W.S. Humphrey, *Managing the Software Process*, Addison-Wesley Publishing Co., Reading, Mass., 1989.

[2] Motorola Corporate Employee Communication, "Software Solution—Motorola's Strategy for Becoming the Premier Software Company," *Software Quality Matters*, newsletter of the Software Quality Institute at the Univ. of Texas at Austin, Winter 1993, pp. 16–18.

Chapter 2

You Have to Have Software

"Change in the marketplace isn't something to fear; it's an enormous opportunity to shuffle the deck, to replay the game." —Jack Welch, chairman and CEO of General Electric. [1]

Software is a strategic necessity in many organizations today. By *strategic* we mean that software is intrinsically involved in carrying out the central function of the organization. It is indispensable to moving products and services along the organization's value-added chain.

An airline, for example, is in the basic business of selling seats on a particular flight and date to travelers. A seat is highly perishable. Once a flight takes off with an empty seat, the opportunity to sell it is gone forever. In the 1950s and early 1960s airline clerks used index cards and blackboards. As passenger volume grew, the detail began to overwhelm them. A computerized reservation system became *strategically* necessary.

As systems of this kind came into use, executives perceived that different classes of travelers had different needs. Seats could be sold at bargain prices to those willing to travel at off hours or to make reservations weeks in advance. At the other extreme last-minute travelers would pay much more. Getting the highest price for each seat led to still more complicated reservation systems.

Then, along the way, executives discovered that other perishable commodities, such as hotel rooms, rental cars, and theater tickets, could also be inventoried through these reservation systems. And so on. As the computerized system took over more of the functions of the business, it became ever larger and more complex.

The central function of a combat airplane is to bring an explosive device into contact with a target. In World War II it took an average of 4,500 missions to destroy a bridge. Bridges were hard to hit with freely falling bombs. By the Vietnam war, the number of missions was down to 95 per bridge destroyed. In the Iraqi war the ratio for "smart" bombs was about three missions, one bridge. The targeting systems embodied hundreds of thousands of lines of software. In a modern air force software is essential to its central function.

Software can also be an Achilles' heel. The Gulf War's largest single American casualty toll was attributed by US Army investigators to a "freak software glitch." [2] That was the Scud missile hit on a barracks. Ironically the Huntsville, Alabama, Missile Command, responsible for the Patriot missile defense system, had just fixed the problem. Unfortunately the new software had arrived in Saudi Arabia only the day before the attack. It had not yet been installed.

No lives were lost on January 15, 1990, the day millions of long-distance calls failed to go through. It was the day ordinary telephone users discovered that software underlies the operation of the telephone system. You may yourself have been inconvenienced by that hours-long interruption of service. It was caused by an obscure error in the software.

American President Companies, Ltd., has combined software control with one of the world's oldest industries, ocean shipping. It calls itself an "intermodal container transportation and distribution services company." One mode is containerized cargo transportation around the Pacific Ocean, extending from the Persian Gulf on the west, through East Asia, to the West Coast of the United States.

The other mode is "double stack" train transportation across the United States. It is "the lowest cost form of long-haul land transportation," the company said. The two modes, when effectively coordinated, transport high-value cargoes, such as designer fashions, electronic products, and refrigerated goods, from North Asian ports, for example, to New York 10 to 14 days faster than all-water transport.

During transit the items are tracked by a computerized system centered in San Mateo, California, and Hong Kong, connected by satellite telecommunications. This system enables American President to inform its customers of the current location of each shipment. Moreover, the system produces the entire complex of paperwork needed to transfer shipments from one country to the next and from one mode of transportation to another. It transmits documentation in advance from the San Mateo data center to the US Customs Service data center, thus expediting customs clearance.

Within its own organization the system enables American President to have ships, railway cars, and trucks lined up to keep everything moving. One result: the Pacific fleet averaged 97 percent utilization and the stack-train system, 95 percent.

Competitive advantage

Thinking through the main function of a business so that it can be accomplished more quickly, cheaply, and reliably clearly provides an enormous competitive advantage. But the strategic level is not the only one in which software can provide an advantage. Some activities, such as computerized accounting, have been around so long that almost every organization has them. Other applications have been implemented to a degree here and there, but remain to be fully worked out by most organizations.

For example, consider that old bete noire, management information systems. In management consultant Tom Peters' phrasing, "human hierarchies are merely machines that process and agglomerate information, each level adding a further degree of synthesis." [3] In principle, it seems that software should be able to do something of that sort. It should be able to present predigested information for upper levels of management to act upon. In practice, it has proved difficult to get all the nuances that management needs into a software system.

Fast decision makers often make better decisions than slow decision makers, according to a study by Kathleen Eisenhardt of Stanford University and Jay Bourgeois of the University of Virginia. [4] Fast decision makers, they said, "set up systems to collect a range of information on their businesses and markets constantly, and then make decisions using the data available."

Slow decision makers first analyze a problem and sort out the questions that must be answered. Only then do they go out and look for that information. Of course, all that takes much more time than that required by fast decision makers who use information routinely generated by computers.

Impact of information technology

In the early decades of computer use, the new technology was applied mainly to specific, discrete applications. Recently John F. Rockart and James E. Short of the MIT Sloan School of Management found that software has been impacting the entire organization by

- ❏ Changing the organization's internal structure.
- ❏ Putting greater emphasis on team-based, problem-focused work groups.
- ❏ Facilitating the flow of products or services through value-added chains.
- ❏ Integrating business functions across organization divisions.
- ❏ Managing *organizational interdependence,* a firm's ability to achieve "concurrence of effort" among functional departments, product lines, geographic units, and other subdivisions. [5]

Rockart and Short reported further that some observers felt that "hierarchical organizations are steadily disintegrating—their borders punctured by the combined effects of

❑ electronic communication (greatly increased flows of information),
❑ electronic brokerage (technology's ability to connect many buyers and suppliers instantaneously through a central database), and
❑ electronic integration (tighter coupling of interorganizational processes)."

"In this view," they continued, "the main effect of technology on organizations is not in how tasks are performed (more quickly, effectively, cheaply, and so on), but rather in how firms organize the flow of goods and services through value-added chains."

Michael Hammer calls this process "reengineering work." He contends that "it is time to stop paving the cow paths," referring to the winding streets in some old cities that are inherently better at accommodating meandering cows than modern traffic. [6] In other words, using the latest information technology to mechanize old ways of doing business isn't good enough. It may reduce your costs 10 percent when you need 80 percent.

Instead "we should 'reengineer' our businesses: use the power of modern information technology to radically redesign our business processes in order to achieve dramatic improvements in their performance," he said. By reengineering its accounts payable process, for instance, Ford Motor Company achieved a 75 percent reduction in its cost.

The prospect sounds attractive, but Hammer goes on to point out that "no one in an organization wants reengineering. It is confusing and disruptive and affects everything people have grown accustomed to."

It is difficult to think through the strategic essentials the organization must accomplish. It is difficult to bring people around to the new way. It takes vision and persistence.

Whether we call the new ways "reengineering" or "more accessible information," they ultimately lead to a restructuring of the organization itself. If a human hierarchy is just a way of condensing information for top management, information technology can replace middle management—or at least that function of middle management. Certainly, in recent years we have seen tens of thousands of middle managers enter early retirement. Better information structures are beginning to support team-based forms of organization. There is less emphasis on the familiar hierarchical form.

Whatever the dividing lines may be along which we break out the impact of information technology, software is certainly affecting organizations in major ways. There is no doubt that, if they are to maintain a competitive advantage, organizations must find a way through the software maze.

References

[1] N. Tichy and R. Charan, "Speed, Simplicity, Self-Confidence: An Interview with Jack Welch," *Harvard Business Rev.*, Sept.-Oct. 1989, pp. 112–120.

[2] E. Schmitt, "Flaw in Patriot Missile Gave Earlier Warning," *New York Times,* June 6, 1991, p. A7.

[3] T. Peters, "New Products, New Markets, New Competition, New Thinking," *The Economist*, Mar. 4, 1989, pp. 19–22.

[4] K.M. Eisenhardt, "Speed and Strategic Choice: How Managers Accelerate Decision Making," *California Management Rev.*, Spring 1990, pp. 39–54.

[5] J.F. Rockart and J.E. Short, "IT in the 1990s: Managing Organizational Interdependence," *Sloan Management Rev.*, Winter 1989, pp. 7–17.

[6] M. Hammer, "Reengineering Work: Don't Automate, Obliterate," *Harvard Business Rev.*, July-Aug. 1990, pp. 104–112.

"When we get the software right, we do very nicely."

Chapter 3

Planning Software Development

"The problem of project management, like that of most management, [is] to find an acceptable balance among time, cost, and performance."
—Peter V. Norden [1]

"The earliest efforts at software cost estimation arose from the standard industrial practice of measuring average productivity rates for workers" [2]. For software cost estimation, the productivity rate "for workers" was assumed to be the lines of source code produced per person-month. With this productivity rate in hand, management could figure the required number of people and the schedule.

This simple method resulted in a lot of wrong answers. Software development became notorious for exceeding schedule and overrunning budgets. "We have a perfect record on software schedules," said one battered manager. "We have never made one yet." (So far no one has established a home for battered managers. The industry probably needs one.)

"When I see a major program fall way behind schedule or greatly exceed budget, I just shake my head," a long-time observer of the software "tarpit" said. "I know the people in charge gambled and lost."

To overcome this "software crisis," people developed more complex estimating methods. Most of the methods worked fairly well in the hands of their originators. But they didn't travel well. Learning to use them took much hard application, and for most organizations, the software crisis continued.

Frankly the problem is that developing software is actually a complex process. The first estimating models naturally mirrored this complexity. They also contained complications that were not essential in practical estimating.

It took some time for us to realize that practical workers in software development don't need to know the mathematical niceties that used to fascinate us. They don't need to follow the algebraic derivations in detail. Immersed in your own consuming work, you probably have the time and inclination to acquire only the minimum you need to know to plan software development.

In this chapter we are going to show the progression from estimating production to estimating development to estimating *software* development.

Estimating production

In the production process diagram in Figure 3-1, the fundamental inputs are people and time, and the output is the product. People and time are combined at some level of effectiveness, or productivity, to produce the product. In other words

$$Product = People \times Time \times Productivity$$

This production relationship is the basis for the traditional estimating computations in straightforward manufacturing operations.

Production Process

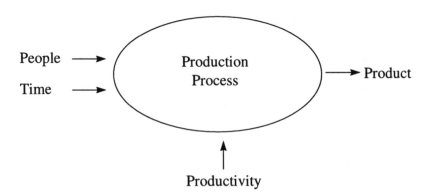

Figure 3-1. Simple production employs people over some period at some level of productivity to produce a product. These are the fundamental elements in estimating production.

The point is how many people are needed over what length of time at the current level of productivity to produce the desired quantity of product. Figure 3-2 diagrams that relationship. In this situation four people are working over seven months, putting in 28 person-months of effort. From experience we know that their productivity is eight units of product per person-month. Total production over the schedule should be 224 units.

When we look at an operating factory in principle, the underlying estimating relationships are this simple. An actual factory usually has complications. The person-hours are not always present because of sickness, strikes, or the inability to hire qualified people. People may be present but idle. When factors like these, sometimes difficult to anticipate in quantitative terms, are taken into account, the simple production equation becomes much more complicated. Production managers have been known to tear their hair out.

Manufacturing Productivity
Productivity = 8 units per Personmonth

Total Production = 224 Units

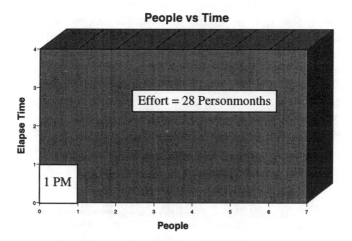

Figure 3-2. People are applied to a manufacturing operation over time at a productivity rate of so many units of product per unit of effort (person-month). Adding up the person-months under the straight line gives a total effort of 28 person-months. This amount of effort at a productivity of eight units per person-month produces 224 units of product.

Nevertheless, at a fundamental level people applied over time at a productivity rate produce a product.

Estimating a development project

Developing a product differs from *manufacturing* a product in several respects. First, development consists of sorting out and solving a series of problems. In repetitive manufacturing most of the problems have already been solved. The effort involved in identifying and solving problems is hard to estimate. Second, the development process begins with a few people, builds up to a peak, and then tails off. A manufacturing process generally operates indefinitely with the same number of workers. If the number changes, the reason is a change in the order rate, not a change in the number of problems to be worked on.

Peter V. Norden of the IBM Poughkeepsie Development Laboratory analyzed the hardware development process in the 1950s: "The succession of purposes with which we work on projects generally are: planning, designing, building and testing a prototype, engineering activities associated with release of the product to plant, occasional redesigning, and a small number of cycles for product support and cost reduction" [1].

Each activity begins with a few key people sorting out the problems to be solved. As they distinguish problems that can be assigned to someone else, they bring in additional people. At some point they begin to find themselves having few new problems and completing work on some of the existing ones. Consequently the number of required people begins to decline. In the meantime, the number of people on the following activity has begun to build up.

Norden discovered that he could represent this type of work pattern with a Rayleigh curve, which has an algebraic equation. With this mathematical framework he was able to project the number of people and time schedule. From these projections he calculated the effort and cost.

Estimating software development

The black box representation of the software development process diagram in Figure 3-3 is similar to the manufacturing representation in Figure 3-1.

In the 1970s Putnam established that the application of people over time to a software project often follows the Rayleigh curve, as Figure 3-4 illustrates. He also discovered that the separate activity curves sum to an overall project curve. This curve, when represented in Rayleigh mathematics, enabled him to project people, time, and effort for the entire project.

Software Production Process

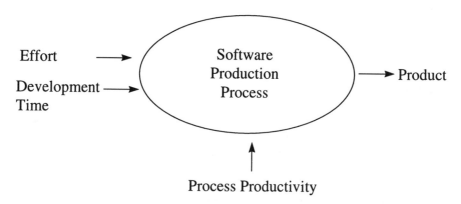

Figure 3-3. For the software development process, a combination of development time, effort, and a measure of the productivity of the development process yields the quantity of product.

Rayleigh Staffing Profile
with sub-phases shown

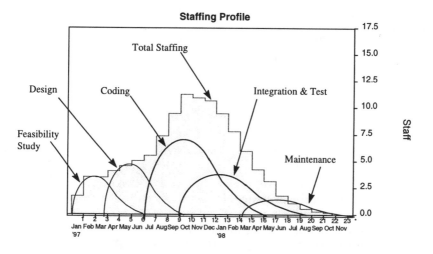

Figure 3-4. The set of small curves represents phases of the overall project. The sum of the effort under the small curves equals the total effort under the large curve. Although managers may not staff according to this curve, inevitably the way work gets done follows this pattern.

Of course, there is nothing magical that *requires* staffing to follow a Rayleigh curve. People can staff a project any way they feel like. If they are lazy planners, they like to level-load. It is easy to build a budget that way. If they are driven by the need for cash flow from the customer, they like to load up fast—often to a high level and level-load after that.

However, regardless of what staffing method a manager uses, something does follow the Rayleigh pattern: *the way the work is done.* Functionality gets created that way. The code gets generated that way. The defects are insinuated that way (since they are proportional to the valid code produced).

Now, if the project manager has a good feel for the way the project is going (during buildup), she senses the need for more people as the team uncovers additional problems, and she applies people that way. (There is some time lag in most cases.) So this—the work—is the real driver. This is what the Rayleigh curve is really portraying. Thus, staffing the Rayleigh way is ideal, most efficient, and most economical.

In contrast, level-loading people on a development project tends to waste some of the staff time. Early in a project, as the team is still uncovering problems, there will be more people than there are problems. The excess people twiddle their thumbs. Later (toward the peak of the Rayleigh curve) when problems are abundant, people are less abundant. The budget for the people then needed has already been expended. Still later, though the number of problems is declining, people (in the sense of people supported by the project budget) are even scarcer.

Since the Rayleigh curve is more elaborate than the level-loading line of the production diagram (Figure 3-2), we are not surprised to find that the input terms of the software production equation are different. Effort (person-years) replaces people. Development time (years) replaces time. And process productivity replaces simple productivity. Thus, the software estimating equation becomes

$$\text{Product} = (\text{Effort}/B)^{1/3} \times \text{Development Time}^{4/3} \times \text{Process Productivity}$$

where *product* refers to the size of the software system in source lines of code (or to function points converted to SLOC) and B is a correction factor for complexity taken from an empirically determined table that is related to the size of the system.

The fractional exponents imply an important message: the software equation describes a far more complex process than manufacturing. The use of relatively simple software estimating methods in the past, based on manufacturing forerunners, goes far toward explaining the dismal record of many software projects.

The presence of four factors in one relationship brings out another important point: the four factors are interdependent. That is, if management action pushes one factor up or down, the other three react accordingly. These effects are a complex subject to which we will return in later chapters.

Process productivity in the equation refers to the effectiveness of the entire project organization, not just to the productivity of the individual programmer. Many factors enter into this organizational productivity: management, supervision, software tools, methods, workstations, and programming language, as well as the skill and experience of the people. In addition, project complexity has a significant bearing on the accomplishment of the work. Therefore, we incorporated all these elements into our term.

We use the term "process productivity" instead of "productivity" to distinguish it from the simple measure of software production—source lines of code per person-month—to which "productivity" has traditionally been applied. It is true that the conventional term is analogous to the manufacturing productivity term, number of pieces per person-month, but the traditional term, like it or not, does not encompass all the factors entering into the effectiveness of a software organization.

As in the hardware development Norden analyzed, people working over time in software development can be depicted by a Rayleigh curve, as Figure 3-5 shows. The area under the curve again is effort, as indicated by the small person-month square. The development time, marked by the vertical line dividing the development period from the operations period, is set at a point called "full operational capability."

Effort after that point is devoted at first to correcting the remaining defects and then to system maintenance. The entire curve represents the "software life cycle."

"The future is uncertain—you can count on it." —Murphy

Projecting the future, or estimating effort and time, cannot be an exact science. The elements entering into the estimating equation are themselves uncertain to various degrees. This uncertainty, or risk, can be visualized in various ways. One is by showing the 50 percent line and the 95 percent line, as Figure 3-5 does. That is, the probability is 50 percent that the team can complete the project along this line. Increasing the effort applied improves the probability of completion, in this case, to 95 percent, as the upper line shows. These lines enable those concerned with evaluating the project to assess the degree of risk in the project they are undertaking.

Still, even though software estimating is not an exact science, this method (based on the software equation and Rayleigh curve) has been in use for almost two decades by more than 300 clients in North America, Europe, East Asia, Australia, India, and elsewhere. Its usefulness has

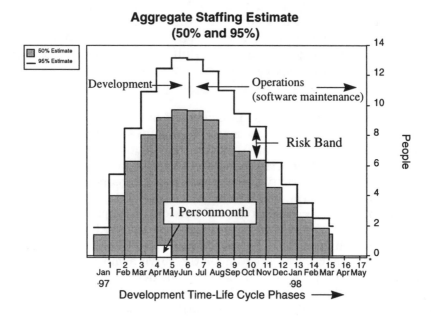

Rayleigh Manpower Curve
Risk interval above estimate

Figure 3-5. The application of people over the software life cycle follows a Rayleigh curve. Development is essentially completed by the time marked by the vertical line, representing delivery time.

been demonstrated in practice thousands of times. There is this "existence proof" that it works. We cite two cases.

Case 1: Single user

A year into a large project, it was not proceeding as planned. Ed Gillingham, an internal consultant at Multiple Systems Corporation, needed an estimate of remaining cost and the date of completion. Using the software estimating method outlined above, he predicted the project would take another 14 months to complete and would run well over the initial cost estimate.

"The answer was not liked," Gillingham recalled. "Project managers thought the prediction wrong."

Months later Gillingham compared the final project outcome with his predictions. "Well, my figures were wrong," he chuckled. "They missed by a month on schedule and one half percent on total cost."

Case 2: Past projects

One way to assess the effectiveness of an estimating method is to apply it retrospectively to completed projects. Here the management numbers are known. You can then compare the known numbers with the numbers the method predicts.

The Electronic Systems Division at Hanscom Air Force Base did just that. It applied this estimating method to 10 major projects for which it had kept very complete data. Captain Roj Karimi, chief of the Cost Analysis Division for Intelligence and C³CM Systems, demonstrated an accuracy of 96.5 percent on schedule and 91 percent on cost. Moreover, the 10 projects were consistent with industry norms on the measure of productivity employed by the method. In fact, they were within the expected range for projects built with modern software technology.

"The study should be of particular interest to acquisition program managers, cost analysts, and other Air Force decision makers who must obtain the software cost information on which their decisions rely," Karimi summed up.

Using the estimates

Estimating the management numbers for software development is merely a logical extension of estimating the corresponding numbers for a manufacturing process or a hardware development project. Perhaps "merely" underplays the difficulty of this undertaking, however. As the processes become more complex, the methodology becomes harder to work out and looks more forbidding when it is found.

Still, on a fundamental level—

❑ There is a mathematical relationship linking certain inputs to the software-development black box to the output, the software product.

❑ There is a mathematical relationship that lets us project people (or cost) over time.

These relationships do solve the estimating problem. You need not concern yourself with the forbidding mathematics. A computer program does the actual computations.* Then it draws curves similar to those in this chapter, enabling you to visualize the project.

---

*The program is SLIM (Software Life-Cycle Model) and related tools from Quantitative Software Management, Inc. You can also calculate the management

The management numbers for *quantity* of time, effort, process productivity, and product size are very important to managers, but the *quality* of the product is just as consequential. In Part IV we explain how you can predict and visualize the management numbers that characterize quality and use them to track reliability.

References

[1] P.V. Norden, "Useful Tools for Project Management," from *Operations Research in Research and Development,* edited by B.V. Dean, John Wiley & Sons, New York, N.Y., 1963.

[2] L.H. Putnam, "A General Empirical Solution to the Macro Software Sizing and Estimating Problem," *IEEE Trans. Software Eng.,* Vol. SE-4, No. 4, July 1978, pp. 345–361.

---

numbers manually using a pocket calculator, as we describe in *Measures for Excellence: Reliable Software on Time, within Budget* (Prentice-Hall, 1992). Moreover, others have developed a few generally applicable software estimating methods, and some are available as commercial programs.

These methods are rather complicated mathematically. Consequently, they are easier to use in the form of a program. As a result we do not describe the methods or programs in full detail. Much of this detail is embodied in a program (where it belongs), not in your head. That the details reside in an application program for personal computers, however, means that its commercial existence does intrude into this book from time to time.

Part II

Pattern of Measurement

To carry on any economic activity you need a pattern of measurement because resources and time available are limited. Some kind of measure of the resources being applied and the time passing ensures that the activity is indeed "economic." A product must be brought to market for a price the market is willing and able to pay. A public service must be accomplished within the immediate limits of a budget and the overall limits set by tax collections. When software development is carried out as a business or governmental activity, it must satisfy these limitations.

In the long run private software managers do discover that they have reached the market too late or at too high a price. Public software managers do discover that they have exceeded budget and schedule. The trouble is this kind of feedback is terribly tardy. In the meantime projects, companies, and people have failed. Funds have been wasted. Needs have gone unmet.

A pattern of measurement enables projects to establish realistic plans and then to gauge where they are against the plan. Managers need feedback in time to do something about problems before it is too late.

That is our objective in Part II. Everyone would probably agree with this goal in principle, but the large percentage of software organizations not defining, collecting, and keeping data implies that something more than a goal is needed.

One difficulty is that scores of metrics have been proposed, but for economic reasons a frugal project must limit itself to a few. So which few? To some extent, these scores of metrics are a smoke screen. In actuality working projects have kept only a few measurements: source lines of code; effort, manpower, or cost; schedule or development time; and number of defects. Consequently, because a pattern of measurement must be derived from the analysis of a database of some size, the pattern must be based on the measurements that have actually been kept.

The pattern we are going to describe in the next set of chapters is a macropattern, covering the entire software development process. Of course, there are specialized parts of the process for which other metrics may be useful. For example, some organizations keep records of the length of time it takes to fix a defect after discovering it. That is a level of detail with which our macropattern does not deal, but it does not prevent you from keeping this record or using other more specialized metrics.

Chapter 4

The Rhythm and Pattern
of Software

*"There is a rhythm and a pattern between the phenomena of nature
which is not apparent to the eye, but only to the eye of analysis."*
— Richard Feynman, Nobel prize-winning physicist [1]

"Real-world problems tend to be enormously complex," observed the
authors of a text on quantitative analysis for business decisions [2]. "In
today's industry, there is no shortage of 'information,' " noted two other
authors, introducing their text on how to extract the main features of
relationships hidden in masses of data [3].

When Johannes Kepler was trying to explain the motion of the
planets, he had the mass of observations collected by the Danish astrono-
mer, Tycho Brahe. The answer was not obvious to his eye. At that point,
mankind had not learned that the laws of physics are fundamentally
simple. He found the answer only after years of analysis.

When you are immersed in day-to-day business operations, you feel
an almost astronomical quantity of detail swirling around you. The purpose of organization, in addition to getting the basic work done, is to help
keep track of all this detail. Even with computers, however, an organization may have trouble staying on top of the detail. It may be lacking a
Kepler to find a pattern in the swirling facts.

Finding a pattern is the business of a Kepler, not of the harried
denizens of an overburdened organization. Just as the original Kepler
had to work with the observations collected by Tycho Brahe, a modern-
day Kepler would have to work with data that someone has collected. In
software, that data has to be the "management numbers" that working
software organizations have recorded. There are only five:

❑ Amount of function created (or/and modified) in a product (often measured in source lines of code)

❑ Schedule (or duration or development time)

❑ Effort (person-months or person-years or people accumulated over time or the equivalent in money)

❑ Productivity or efficiency

❑ Mean time to defect or its reciprocal, defects per month

Management numbers increase with size

At the time managers are planning a software project, they have some idea of its size. They want to project schedule, effort, and defects. It seems logical, therefore, as a first step toward finding some pattern in the software development process, to plot these three management numbers for completed projects against size.

Estimators first made plots of this kind in the 1970s from the management numbers of a few hundred projects. By 1994 Quantitative Software Management had assembled more than 3,885 projects totaling more than 300 million source lines of code, or three million function points, in over 200 languages. These projects embodied 65,000 person-years of effort. We are adding 200 to 400 projects each year, keeping the database up to date with current methods.

The length of schedule increases with size, as Figure 4-1 shows. Similarly, the amount of effort grows with size, as shown in Figure 4-2. The number of defects also rises with size, according to Figure 4-3.*

These scatter plots reveal two important patterns. First, although the numbers of interest increase with project size, they do not increase in direct proportion to size. They increase at different rates. Second, there is a large range of values at each system size.

---

*All three of these diagrams represent all the projects in the QSM database. We call it a "mixed application database" because it contains all the different kinds of applications. Figure 4-3 (defects) contains fewer points than the time and effort figures because fewer organizations have recorded error data.

Also, we have drawn both the horizontal and vertical axes to logarithmic scale. Log scale lets us represent an enormous range of project sizes on one small diagram. In these figures, for example, project size ranges from a few thousand lines of source code to over one million lines. Log scale also turns nonlinear behavior, which would appear on diagrams as curved lines, into straight lines.

Schedule Behavior

QSM Mixed Application Data Base

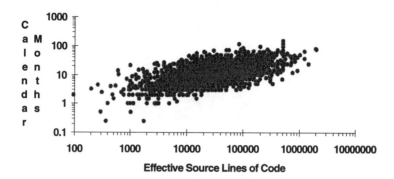

Figure 4-1. The larger the project (in source lines of code), the more development time it takes. There is a great deal of variation in development time at each project size.

Effort Behavior

QSM Mixed Application Data Base

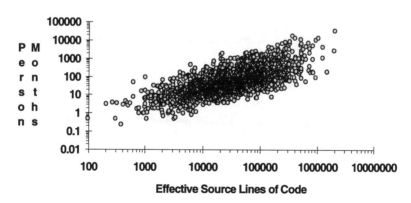

Figure 4-2. Effort increases with growth in project size much more rapidly than development time does. There is much variation at each size.

Defect Behavior
Total Found from Integration to Delivery

QSM Mixed Application Data Base

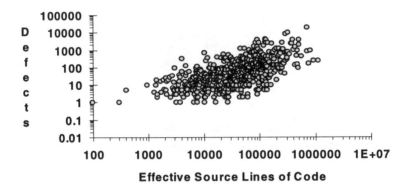

Figure 4-3. Defects increase dramatically with size growth. The number of defects varies widely at each size.

A closer look gives some interesting insights. An average project with a small number (10,000) of source lines of code appears to have a development time of about 10 months, effort of about 20 person-months, and about 20 defects (defined as those discovered between system integration test and delivery).[1]

A larger project at 100,000 SLOC has a development time of 20 months, effort of 200 person-months, and 200 defects. Thus, for a size increase of 10 times, the three management numbers increase by 2x, 10x, and 10x respectively. We repeat these numbers in Table 4-1.

While all three management numbers are increasing, they are expanding at different rates. A moment's thought suggests one answer. As projects become larger, we put more people on them, reducing the calendar time larger projects might otherwise take. Thus, calendar time increases more slowly than effort, as Table 4-1 shows.

---

[1]We eyeballed these values off the three diagrams at the center of mass of the dots at 10,000 SLOC.

Table 4-1. Under the circumstance listed in the first column, development time, effort, and defects vary as shown.

Circumstance	Development Time (months)	Effort (person-month)	Number of Defects
For a project at 10,000 SLOC	10	20	20
For a project at 100,000 SLOC	20	200	200
For a size increase of 10x	2x	10x	10x

At the moment the foregoing observations are merely examples of what a database can reveal. In Parts III and IV, we show how management can put them to good use in trading off one number to gain an advantage in another number. We shall find, for example, that lengthening development time a few weeks or a few months at the time of planning a project will reduce the overall effort and cost.

Source lines per person-month is a poor metric

Effort over a period of development time at some means of measuring productivity results in a product, as Figure 3-3 illustrated. For estimating, we can reverse this process. An estimate of product size divided by a measure of productivity provides an estimate of the effort to build the product:

$$\text{Effort} = \text{Product/Productivity}$$

Or, if we measure effort in person-months, product size in source lines of code, and productivity in SLOC/PM:

$$\text{Person-months} = \text{(Source lines of code)/(SLOC/PM)}$$

To implement this simple estimating formula, many organizations have used source lines per person-month, or SLOC/PM, as their measure of the productivity of software development. Unfortunately, it is an uncertain trumpet! Using it as the principal basis for estimating project effort and the schedule over which effort is to be expended has resulted in highly inaccurate projections.

Some observers feel that SLOC is a measure only of code output. It is not, in their view, a measure of software production elements such as requirements, specifications, design documents, test plans, trouble reports, user manuals, and other documentation. In addition to coders, there are managers, systems engineers, analysts, designers, testers, writers, and support staff. Coding may occupy as little as 10 percent of project person-hours.

We view SLOC, however, as a measure of the ultimate output of the software process. We see the preceding activities as something that must be carried out to produce code. We see inspection and testing as necessary to get the code right. We view SLOC as the product that gets compiled into executable code, goes into the computer, and does the work the user wants. On this basis, SLOC is a viable measure of the amount of functionality in a software product. Moreover, it is usually the *only* measure available on working projects.

At first glance, the person-months in SLOC/PM appear to be a rather solid metric. As we demonstrate later, however, the amount of effort required by a given project varies quite widely with the length of development time allowed. In brief, if the customer and project management allow a reasonable development time, as opposed to a very short period, the amount of effort (person-months) required to complete the project is much less. Thus, depending on the way an organization has carried out past projects, the effort factor in this conventional productivity formula may be quite uncertain.

The variability of these two metrics, SLOC and PM, is apparent in Figure 4-2. This diagram plots person-months of effort against size measured in SLOC. The amount of effort at any given size varies widely. For example, at a size of 100,000 SLOC, the person-months of effort range from about eight to 3,000. The corresponding productivity in SLOC/PM would range from 12,500 down to 33. The wide dispersion of the circles, representing actual projects, at all sizes, indicates that SLOC/PM is a slippery measure for estimating effort and schedule.

Many organizations have not realized that this conventional measure of productivity is so variable. They have based a productivity rate on a few past projects, not realizing the degree to which this measure varies. Then, when they estimate another project, they are surprised that it turns out poorly.

Our conclusion, then, is unequivocal: *Do not use SLOC/PM as an indicator of productivity for planning and estimating.* Too many hidden (but explosive) factors are at play. Shortly we develop a better measure, the "process productivity" measure introduced in the last chapter.

Management numbers vary widely at same size

The development time, effort, and defects recorded for projects of about the same size vary enormously. For example, at 80,000 SLOC, development time varies from three to 100 months, as Figure 4-1 shows. Effort ranges from eight to 2,000 person-months, as Figure 4-2 shows. And the number of defects between system integration test and first operational capability extends from 10 to 4,000, as Figure 4-3 shows. Table 4-2 summarizes these ranges.

In other words, if we were trying to estimate the duration of a project expected to be 80,000 SLOC, we could project a number only in the range of three to 100 months. If we weren't sure of the size estimate, guessing that it might fall between 60,000 and 100,000 SLOC, for example, the vertical dispersion on the scatter plot would be even greater.

Let's consider the possible causes of this dispersion. We know as a matter of practical observation that some kinds of software are more difficult to design and code than others. Real-time embedded programs, for example, take more effort than business systems of the same size. It is likely, therefore, that this "complexity" is one cause of the dispersion.

We also know that some project organizations work more effectively than other organizations. Most of us believe that quality of management and people, high-level languages, good programming practices, software tools, workstations, and other aspects of the development environment affect organizational effectiveness.

In the early 1970s C.E. Walston and C.P. Felix at IBM analyzed 68 factors they thought might affect productivity. Twenty-nine of them "showed a significantly high correlation with productivity" [4]. A few years later Barry Boehm selected 15 factors for use in his Cocomo estimating model [5]. In effect, studies such as these confirmed our collective intuition: Many factors affect software development.

Table 4-2. The management numbers vary enormously at a given size, as the broad vertical dispersion of the circles on the scatter plots demonstrates.

Circumstance	Development Time (months)	Effort (person-months)	Number of Defects
For a typical project at 80,000 SLOC, the range is	3 to 100	8 to 2,000	10 to 4,000

Furthermore, we know that some organizations have improved their ability to develop software over the years. And a few have deteriorated! Consequently an estimate of productivity that might have been valid in 1985 might be off the mark in 1990.

It seems likely that the vertical dispersion on the scatter plots is the result of some combination of project complexity, organizational effectiveness, and time period in which the evaluation was made. Thus, our next task is to examine that complexity.

References

[1]   R. Feynman, *The Character of Physical Law*, The MIT Press, Cambridge, Mass., 1967, 173 pp.

[2]   H. Bierman, Jr., C.P. Bonini, and W.H. Hausman, *Quantitative Analysis for Business Decisions,* Richard D. Irwin, Inc. Homewood, Ill., 1973, 527 pp.

[3]   N.R. Draper and H. Smith, *Applied Regression Analysis*, John Wiley & Sons, Inc., New York, N.Y., 1966, 407 pp.

[4]   C.E. Walston and C.P. Felix, "A Method of Programming Measurement and Estimation," *IBM Systems J.*, Vol. 16, No. 1, 1977, pp. 54–73.

[5]   B.W. Boehm, *Software Engineering Economics*, Prentice-Hall Inc., Englewood Cliffs, N.J., 1981, 767 pp.

Chapter 5

The Pattern of Complexity

"Complexity is a not-so-warm feeling in the tummy."
—Bill Curtis, as quoted by Tom DeMarco [1]

"The nature and effects of complexity have been studied for years by systems people," says Tom DeMarco, a well-known software consultant, "but our industry has not even been able to settle on a definition."

It would seem that just what complexity is and what effect it has on management numbers is a little vague. Nevertheless, most observers believe, in the words of one of them, Edward T. Chen, "We have known for a long time, based on our experience, that a complex program takes more time to develop" [2]. In an experiment Chen found that productivity is "strongly related to the program's complexity measure," referring to a measure of complexity that he developed.

Aspects of complexity

Walston and Felix, mentioned in Chapter 4, found that of their 29 factors that "correlate significantly with programming productivity," four reflect some aspect of complexity:

❏ Customer interface complexity
❏ Overall complexity of code developed
❏ Complexity of application processing
❏ Complexity of program flow [3]

Similarly, one of Boehm's cost-driver attributes, "software product complexity," is second only to his "personnel/team capability" attribute in its effect on estimating effort and time [4].

So, we may not have a good definition of complexity. As the judge said about pornography, "I can't define it, but I know it when I see it." When we see complexity, we know it has some relationship to how much time and effort the project will take.

Unfortunately, without some concrete measure of the complexity inherent in a project, software people have had no good basis for estimating its effect on schedule, effort, and cost. We know it has an effect. The issue becomes, if you are going to produce estimates within engineering limits of accuracy, how much effect?

What the database shows

Our 3,885-project database classifies each project into one of nine application types, as listed in Table 5-1. As a first approximation, application type appears to be related to complexity.

In the previous chapter we plotted the management numbers of systems of all types in the database against size. In this chapter we sort out the systems by application type. In Figure 5-1, for example, we plot effort recorded for business systems against the corresponding size. In Figure 5-2 we plot effort for real-time systems against size. If you look closely at the two figures, you can see that the real-time systems take more effort on the average at each size than business systems.

Table 5-1. Organizations reporting data to us classified their projects under nine application types. Most of the reports are on business systems, the most common type of project.

Business	66%
Real-Time Embedded	7%
Scientific	7%
Systems Software	6%
Telecommunications	5%
Command and Control	5%
Avionics/Space	2%
Process Control	1%
Microcode/Firmware	1%

Effort vs. Size
Business Systems

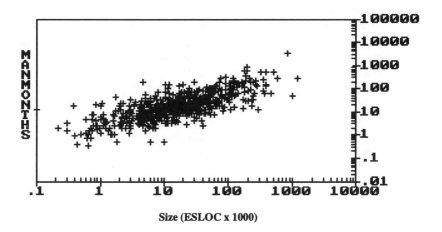

Size (ESLOC x 1000)

Figure 5-1. We have sorted business systems from the entire database and plotted them separately.

Effort vs. Size
Real Time Systems

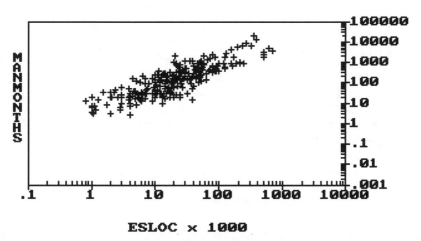

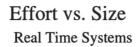

Figure 5-2. When we plot real-time systems separately, the effort recorded at each size is greater than business-systems effort at the same size.

In Figure 5-3 we dispense with the hundreds of little crosses and instead draw in a line representing the mean value of the crosses. This line shows the average effort at each size. The lines above and below the mean line represent one standard deviation above the mean and one standard deviation below the mean. About 68 percent of the data points—the bulk of the cases in this sample—lie between the standard-deviation lines. Project data for only about 16 percent of the cases is either better or worse than the projects included between the lines. This representation gives us a way to focus on the typical projects.

The next step is to draw the mean lines for several of the application types on one figure, Figure 5-4. As the figure shows, the mean lines constitute a set of parallel lines on the log-log field of effort versus size. In other words, we have "stratified" the database by application type. We can draw similar diagrams for development time and number of defects.

Effort vs. Size

Command & Control System Trend Lines

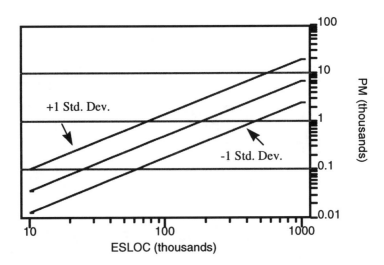

Figure 5-3. The mean line and the two standard-deviation lines, which we may call the trend lines, are another way to represent the scatter plots. This diagram represents the command and control application area, which differs in productivity from the business systems and real-time systems shown in the preceding diagrams.

Effort vs. Size
Trend Lines for Several Application Types

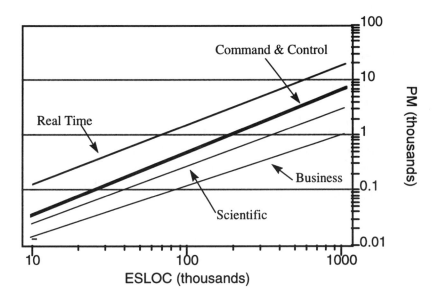

Figure 5-4. When we stratify the database by application type, we see that application type (or complexity) is one factor causing the vertical spread of the amount of effort at each size.

In estimating, if we knew the application type, we could go to that line and read off a more accurate estimate of effort than we could off the mean line of all the applications together. Of course, there is still a vertical dispersion of the effort values above and below the mean line, as the standard deviation lines of Figure 5-3 show.

Application type, of course, is only an approximate stand-in for complexity. If we could measure complexity more precisely, we could draw many more lines on a figure, such as Figure 5-4. Most of the approaches used to assess complexity, however, depend on ratings that are based on the judgment of managers.

Rating scales generally have five or six gradations. A manager's judgment may easily be off one gradation. In percentage terms a given rating is about 20 percentage points wide and might be off another 20 points. That much potential error seriously distorts an estimate.

Of course, the future is notoriously difficult to nail down. Most people expect estimates to be inexact, but a method that builds in up to 40 percent error in the complexity factor suggests that it is desirable to look further.

We need a factor that is based on counting or measuring, not fallible human judgment. We need a factor that reflects conditions in an organization at the present time, but can change as circumstances change. We turn to this need next.

References

[1] T. DeMarco, *Controlling Software Projects*, Yourdon Inc., New York, N.Y. 1982. 284 pp.

[2] E.T. Chen, "Program Complexity and Programmer Productivity," *IEEE Trans. Software Eng.*, May 1978, pp. 187–194.

[3] C.E. Walston and C.P. Felix, "A Method of Programming Measurement and Estimation," *IBM Systems J.*, Vol. 16, No. 1, 1977, pp. 54–73.

[4] B.W. Boehm, *Software Engineering Economics*, Prentice-Hall Inc., Englewood Cliffs, N.J. 1981, 767 pp.

Chapter 6

Calibrating the Software Process

Calibrate: To determine the correct range for an artillery gun or mortar by observing where the fired projectile hits.
—Random House College Dictionary

If the software estimate were a projectile, calibration would determine where it hits—over or under schedule or cost.

In Chapter 3 we surmised a relationship between the management numbers, based on the analogy to production estimating formulas, of the following nature:

$$Size = Effort \times Development\ Time \times Fourth\ Factor$$

Three of these four elements—size, effort, and development time—are subject to measurement, basically by counting source lines of code, person-months of effort, and months of time. Of course, there are some problems in standardizing these terms and more problems in setting up organizational discipline to enforce the standards. But those things can be done. In principle, we can count these three factors quite precisely.

The fourth factor is another story. In elementary algebra it would be the "constant." It is perhaps comparable to the multiplier based on attributes—experience, tools, methods, environment, and so on—in estimating procedures such as Cocomo.[*] For the moment we leave it undefined. We'll just call it the fourth factor.

---

[*] The Cocomo model uses two equations:

$$Effort = A*(KSLOC)^b * (F1*F2*...*F15)$$

$$Development\ Time = C\ (Effort)^d$$

Fourth factor

As we showed in the last chapter, we can stratify the scatter plot of effort versus size by application type. With nine application types we would obtain nine mean lines. Consequently, a prediction of effort based on a stratified mean line would be much better than a prediction taken off the mean line of all the applications in the database. In effect, we need to find a way to draw a large number of these stratified mean lines.

Because nature, at least in the physics realm, seems to work with simple equations, Galileo found only one gravity acceleration constant. In software development, however, our scatter plots imply that the fourth factor is not one number. It is a range of numbers. Software development seems to be more messy than the law of gravity. Therefore, we had to work with a larger database than Galileo needed.

Some years ago when our database contained several hundred completed projects, we had values of size, effort, and time for these projects. When we rearranged the foregoing equation by algebraic methods,

Fourth Factor = Size / (Effort × Development Time)

we obtained about 20 values of the fourth factor. These values ranged from less than 1,000 to more than 46,000.

Generally, organizations do not count size, effort, and time that closely. In many cases projects report size only to the nearest 5,000 or 10,000 source lines of code—85,000 for a product that is actually 81,587 lines. That is an error of 4.2 percent.

Similarly, they may round off effort to the nearest 10 person-months. Even if projects reported it to the precise person-month, they would still have difficulty defining just who to count or what hours to exclude from official working time. We discuss the standardization of measurements in Chapter 10.

Development time is the worst of the three. Most organizations count it only to the nearest month. On an 18-month project, the time report might be off by five percent. Development time is also difficult to define. Just when does a project reach "full operational capability"?

Thus, what organizations have recorded in the past depends in large part on how carefully they have defined these terms and the discipline with which they have collected data. Much historical data is imprecise, perhaps in the 10 percent range. In the future, with precise definitions and a good collection discipline, you might achieve one percent precision in your own calibration data.

In consequence, we initially established 18 index numbers to represent the fourth factor. Several years later some of our clients began to exceed index 18 and we extended the range to 25. In 1989 a few organizations beat 25. Table 6-1 shows the index numbers now established.

Table 6-1. The fourth factor. This factor corresponds to the overall efficiency or effectiveness of an entire software organization. Each index number stands for a productivity parameter, the value used in computations. The application types are listed opposite their mean values. The standard deviation of the application types is given in index numbers.

Productivity Index	Productivity Parameter	Application Type	Standard Deviation
1	754		
2	987		
3	1,220		
4	1,597		
5	1,974		
6	2,584	Microcode (6.6)	±2.6
7	3,194	Avionic (7.0)	±3.2
8	4,181	Real-Time (7.5)	±3.6
9	5,168		
10	6,765	Command and Control (9.9)	±3.9
11	8,362	Telecommunications (11.0)	±3.5
12	10,946	Systems Software (11.9)	±4.1
		Scientific Systems (12.1)	±3.9
		Process Control (12.1)	±3.3
13	13,530		
14	17,711		
15	21,892		
16	28,657	Business Systems (16.9)	±4.9
17	35,422		
18	46,368		
19	57,314		
20	75,025		
21	92,736		
22	121,393		
23	150,050		
24	196,418		
25	242,786		

26	317,811		
27	392,836		
28	514,229		
29	635,622		
30	832,040		
31	1,028,458		
32	1,346,269		
33	1,664,080		
34	2,178,309	Highest value reported	
35	2,692,538		
36	3,524,578		
37	4,356,618		
38	5,702,887		
39	7,049,156		
40	9,227,465		

In computations we now carry the index numbers to the nearest tenth, such as 28.4. Thus, the entire range of the fourth factor, from 0 to 40, embraces 410 increments. This number of increments provides considerable sensitivity in computations, perhaps more than the precision of the underlying data warrants.

So far we have been calling this constant in the software equation "the fourth factor." This nondescriptive name underlines the point that we haven't had to know just what it is to establish its values by calibration. By analogy to the production-estimating equations, it is some kind of productivity. By analogy to estimating systems based on assessing software process attributes, it embraces all the 15, 20, or 29 factors that investigators believe have a substantial relationship to productivity.

A larger concept of productivity

The important factors in software development, we believe, can be summarized as

- ❑ The influence of management (investment and process improvement)
- ❑ Software development methods, including programming language in use
- ❑ Software development tools, techniques, and aids
- ❑ Skills and experience of development team members
- ❑ Availability of development computer(s)
- ❑ Application type, and complexity.

These factors comprise a larger conception of productivity than conventional productivity (SLOC/PM) does. To distinguish the fourth factor from conventional productivity, we called it "process productivity." The numbers from calibration then become the *process productivity parameter* and the index becomes the process productivity index, or *productivity index*.

The PI is linear. In truth, however, it represents the exponential scale of the process productivity parameter, as Figure 6-1 demonstrates. In fact, there is a multiplier of 322 times from a PI of 1 to a PI of 25. The multiplier from PI 1 to PI 34, the highest value yet encountered, is 2,889. That is why a difference of one digit in the productivity index makes a considerable difference in the estimates of effort and time.

In a typical development, for example, an organization having a PI one digit above the average completes a project in 10 percent less development time at a cost approximately 30 percent lower. Likewise, with a PI one digit below average, the time and cost are larger by 10 percent and 30 percent respectively.

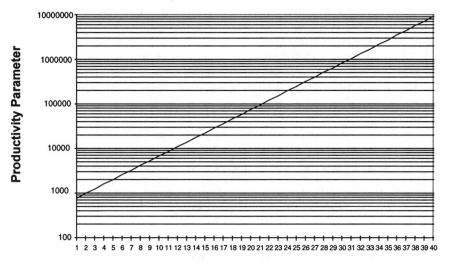

Productivity Parameter vs. Productivity Index

Figure 6-1. When using the PI, keep in mind that the rate of growth of the corresponding process productivity parameter is actually exponential. With the values of the process productivity parameter expressed on a logarithmic scale, the relationship to the PI appears as a straight line.

Low values of the index, relative to the mean of the application type, generally characterize an ineffective software organization. An organization doing complex work, such as real-time embedded code, will have a lower index, other things being equal, than an organization doing less complex, better understood work, such as business systems.

Calibrating your own index

You can calibrate your productivity index from data on your completed projects. For example, suppose Table 6-2 contains the size, development time, and effort of three of your recently completed systems—a large one, a medium one, and a small one.*

The values of the productivity parameters for these three projects are somewhat different, but the nearest productivity index is 12 in all three cases. The fact that three projects in the same organization have the same productivity index is not surprising. Presumably the three groups were working in nearly the same environment.

Still, three groups in the same general environment may differ in their microenvironments. Skills and experience may differ, for instance. There may be small differences in complexity even among apparently similar projects. A higher level manager should not be surprised if one group has a lower PI than another.

The PI represents the ability of your organization to develop software. It also incorporates the complexity of your particular work. Moreover, because it is calibrated from completed projects, it represents your situation at a period when the projects were in progress.

Table 6-2. Calibrating large, medium, and small systems completed by the same organization resulted in the same productivity index.

System category	Size (SLOC)	Development Time (years)	Effort (person -years)	Productivity Parameter	Productivity Index
Large	100,000	1.75	32.0	10,941	12
Medium	40,000	1.20	8.3	10,781	12
Small	15,000	.78	1.1	10,949	12

---

*We used a computer program to calculate the table values. The computations involve several complications that we have spared you. Our book, *Measures for Excellence*, explains these computations more fully, but it is easier yet to rely on the programs produced by Quantitative Software Management, Inc.

If you calibrate on last year's projects, for example, you have a current basis from which to project next year's estimates. Next year you can calibrate on this year's projects and have an up-to-date basis. If your organization is getting better, your PI is always a current basis for new estimates.

And that suggests another use of the PI. When looked at over time it becomes a measure of whether your process productivity is improving, a subject to which we devote Part V.

Chapter 7

Valuing Attributes by Calibration

Most software development models...require calibration, that is, determining coefficients and constants from historical data gathered in a specific environment. —S.D. Conte, H.E. Dunsmore, and V.Y. Shen [1]

The "prim and proper" way to value the attributes used in software-estimating systems is for managers or estimators to sit down and evaluate each factor. For example, "prim and proper" would have you use 20 attributes of the software process, such as personnel experience and tool use, in making estimates. You would then divide each attribute into five levels, each described by a paragraph of text. Next you would assign each level a weight from 0.70 to 1.50. A rating of 0.90, for example, represents better than average experience. A rating of 1.00 is nominal.

When you multiply the weights of all levels of the 20 attributes, you get an overall multiplier. Then, after you initially figure out the amount of effort (or cost), you apply the multiplier to it to allow for the effect of these 20 attributes. Suppose the overall multiplier is 0.85 because you feel you had pretty good people and tools. The final effort estimate would be 85 percent of the initial effort estimate. With good people and so on, you figure you can do the job in fewer person-hours.

In contrast, there is a "bootleg" method of valuing attributes. A friend of ours, Phil, a software manager at the Decorous & Dapper Systems Company, told us about it:

Last week my friend John [also a software manager at Decorous & Dapper] was trying to estimate a new project.

"I just don't know enough about these attributes to rate them at all accurately," John complained. "I've worked in this one company since I left college. I've been with this group for four years. I think they're pretty good, but how do I know how they compare with other software developers? How do I know how our tools stack up with the other tools

available? I always feel so inadequate when I try to pick the closest ratings out of a hundred possibilities."

"I learned something in sophomore physics laboratory that maybe you missed," I said.

"Of course you did. We went to different schools," he replied. "What are you talking about?"

"We had a series of lab experiments to verify the basic laws of physics," I explained. "One of them involved replicating Galileo's demonstration that all falling bodies, regardless of their weight, increase their speed at the same rate."

"Yeah, I learned that in kindergarten," John said. "The teacher dropped an orange and a tennis ball…"

"Well, this was in college and we were a bit more sophisticated than that. We were supposed to derive the constant for the acceleration due to gravity. We rolled balls of different weights down a smooth slope. A ball went over a lever at the top of the slope and another one at the bottom that started and stopped a timer. Given the time and the distance, we were supposed to calculate the acceleration constant.

"The trouble was," I went on, "the answer always came out wrong. Of course, we knew the right answer, 980 centimeters per second per second. Galileo had found it about 400 years ago. The distance seemed right. We could recheck that ourselves. But the time wasn't right—probably something wrong with the timer. My little group took the known distance and the known constant of acceleration and, working backward, calculated the time."

"Good for you," John said. "Today you would get in trouble with the thought police—faking data. One of the things Robert Fulghum [author of All I Really Need to Know I Learned in Kindergarten [2]] learned in kindergarten was 'Don't take things that aren't yours.' "

"Criticism noted," I said smartly. "Someone, not me, later pointed out the deficiency to the professor in charge and eventually they fixed the timer. Meanwhile we had to turn in lab notebooks and pass the course."

"I think it has twisted your character," John mused. "I've noticed that you always vote for the wrong man for President."

"The following year we had a course in instrument design," I continued, ignoring this slur. "I learned that what we had been innocently doing was called calibration in instrument circles. If you make a temperature-measuring transducer, for instance, you don't know what current output corresponds to 0 degree Celsius and 100 degrees Cel-

sius. You stick your device in ice water and measure the current at 0 degree; you stick it in boiling water and mark 100 degrees."

"Very shrewd," John said. "I hope you didn't burn your fingers."

"There is an important principle here, if you'd care to pay attention," I insisted. "When you can't find something you need to know by analytical means, try finding it experimentally."

"That is exactly what I do around here," John said wryly. "When I can't get a good estimate from the 'prim and proper' formula, I get it the hard way. I wait until we complete the project. Then I see how much it cost, how long it took, and how big it is. There's just one little problem with this 'experimental' approach. I get the answers too late. We've been losing money on poor cost estimates and losing customers on late deliveries."

"Ah, you still don't get the point," I gloated. "The money lost on those past projects is regrettable, but what's done is done. They are now history, but they are also completed experiments. You know the effort, schedule time, and size on each one. From those knowns you can now work backward to find out what the multiplier was.

"If you have three or four old projects," I continued, "you can calculate the multiplier three or four times and average them. Because you are going to do your next project with pretty much the same people, tools, and environment, the multiplier is going to be the same, too. Your next estimate ought to be pretty close to the eventual result."

John hesitated for long seconds while emotions passed over his face. "I'm trying to look at your calibration idea from every angle," he finally said. "I can see that calibration gives you the overall multiplier, but how do you go from there to rating the 20 attributes?"

"I rate each attribute as accurately as I can," I replied. "I have the same trouble you do. I don't know everything about all those 20 areas. At any rate I multiply the 20 weights together and compare the rating result with the calibration result. At this point I know the calibration multiplier is the more accurate one, because it is based on solid, countable facts—the size, effort, and time of recently completed projects. Then I go back to the ratings and adjust those I am less sure of up or down one level until the ratings multiplier matches the calibration multiplier."

"They used to call me Honest John in kindergarten," he said. "I knew they were kidding me, but I kind of liked it anyway. It doesn't seem quite honest to fiddle with ratings that you have already made as honestly as you can."

"If you had complete knowledge of the 20 attributes, your honest rat-ings also would be accurate ratings. But your knowledge not only isn't complete, it's really meager. Your knowledge of the figures you use in calibration is complete. Moreover, the figures are accurate, or reasonably so. So your estimate is pretty accurate, and that is honest."

"You sound convincing," Honest John finally conceded, "but you're too crafty ever to be called Honest Phil. Well, I'll try to explain it to my boss. I just hope he had some defective apparatus in college."

We have simply taken Phil's method to its logical conclusion: the fourth factor. We make no bones about its inherent honesty. It is not necessary to break it down into 20 attributes, defined by 20 more or less vague paragraphs. We find it directly by calibrating experience.

Besides, it is easier and more accurate to calibrate a process than it is to analyze it.

References

[1] S.D. Conte, H.E. Dunsmore, and V.Y. Shen, *Software Engineering Metrics and Models*, The Benjamin/Cummings Publishing Co., Inc., Menlo Park, Calif., 1986, 396 pp.

[2] R. Fulghum, *All I Really Need To Know I Learned In Kindergarten*, Ballen-tine Books, New York, N.Y., 1989, 196 pp.

Chapter 8

Estimating Size

Little did the ancient Greeks realize that the universe actually consists of earth, fire, water—and software.
—Norman R. Augustine, CEO, Lockheed Martin [1]

"You can't measure pure thought stuff," sniffed the young programmer, loosening the laces of his tennis shoes to let more blood flow to his brain.

"Maybe not," grunted the old timer, *"but we used to weigh the box of punch cards containing the program."*

"But the program was the holes," the young programmer shot back. *"They don't weigh anything."*

If you could weigh-count punch cards in the old days, you can count source lines of code today—after you complete the program. The issue for software management is how to estimate source lines of code early enough to be of use in forecasting schedule and effort.

We are using "source lines of code" here as a stand-in for the amount of work to be done. If we could directly estimate the amount of work embedded in the function to be created, forecasting schedule and effort would be easier. We do work to create the function, which the programmer expresses in source lines of code. Of course, the amount of work to be encountered in any intellectual activity, including software development, cannot be known precisely before we undertake the task. As a practical matter, however, organizations have been counting source lines of code in all kinds of software projects for several decades. We have a body of experience to work with.

The only other measure of size that has seen much use is function points. Function-providing elements include inputs, outputs, master files, inquiries, and interfaces. They are a user view of software functionality. As soon as the user has established what the software is to do, project members can count these function-providing elements. Then they can determine the function-point count and convert it to source lines by established methods [2].

The main ideas

Estimating size is the heart of the software-project estimating process.

Start by selecting a few people with the best experience in the application area of the proposed project. The key steps in software sizing are not a matter of formulas you can run on a computer. They are a matter of judgment. The best judgments come from good people.

Ask them to estimate size on the basis of the information available. Early on, with little information known, estimates will be of the "ball park" variety. Later on, after a feasibility study or functional design, when they know more about the project, knowledgeable people can refine the estimate.

Suggest that they nominate a range of sizes—broad at first, narrower as the project becomes better defined. The method we suggest picks three points:

❏ The minimum possible size.
❏ The most likely size.
❏ The maximum possible size. Select the range so that there is a 99 percent probability that the eventual actual size falls within it. In other words, only a few outlying possibilities should be less than the minimum or more than the maximum size.∗

Have them combine their separate three-point estimates by established statistical methods into a single estimate, plus and minus a range. An estimate obtained by statistical combination will turn out to be closer to the ultimate actual value than any one estimate by itself. "The more pieces of software we estimate, the more we get the law of large numbers working for us to reduce the variance of the estimate," Barry Boehm observed [3].

∗ The details of these statistical methods are in *Measures for Excellence: Reliable Software on Time, within Budget* (Prentice-Hall, 1991), as well as in standard statistical textbooks. The methods are also incorporated in QSM's computer-based product, Size Planner.

The outcome of the statistical approach is a single number. Because it is an estimate, not a certainty, the method shows the degree of uncertainty of this size estimate by another number, called the standard deviation. It signifies a 68 percent probability that the eventual actual size will fall within plus and minus one standard deviation of the estimate. Once you express the estimated size in this standard way, extensive analytical techniques become available. For instance, you can weigh the degree of risk in the project, a topic we pursue in Chapter 13.

You know as a matter of common experience that a range is the reality, not some exact number like 88,765 SLOC. What you need is a method to quantify this margin of error and then to deal with it. This set of ideas provides that.

Beyond the basic estimate

"In 1798, Eli Whitney contracted to deliver 10,000 muskets to the Continental Army within 28 months," Augustine noted. "As things worked out, they delivered them in 37 months, or in about one third more time than had been anticipated" [1].

Almost two hundred years later major new systems delivered to the US military by large industrial companies also exceeded schedule by about one third, Augustine added. For example, in 1986 two studies for the President's Blue Ribbon Commission on Defense Management found that "on average, cost overruns on selected major weapon systems were about 40 percent" [4].

In 1992 Quantitative Software Management found the pattern of schedule slippage summarized in Figure 8-1 and effort overrun in Figure 8-2. In 664 reports of schedule slippage, 71 percent were within 130 percent of their planned schedule. That means that 29 percent, almost one third, exceeded their planned schedule by one third or more. Similarly, in 622 reports of cost overruns, 77 percent were within 130 percent of their planned cost. That means that about one quarter exceeded their planned cost by one third or more.

Butler Cox plc (now CSC Index), a London-based international management consulting group, found that the average effort overrun was 37 percent of the planned effort on 344 sizable systems-development projects. The schedule slippage was 32 percent of the planned development time [5].

The accounting firm, Peat Marwick Mitchell & Co., surveyed 600 of its largest clients. About 35 percent had major runaways of cost or schedule. The runaway problem was so common that Peat Marwick set up a group to rein them in. In its first two years of operation the group racked up $30 million in fees from 20 clients [6].

Schedule Performance

Percent of Systems
with Schedule Slippage

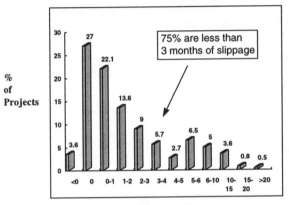

Months of Schedule Slippage

Figure 8-1. If we accept two months slippage as approximately meeting schedule, then one third (33.6 percent) substantially exceeded the planned schedule.

Effort Performance

Percent of Systems
with Effort Overrun

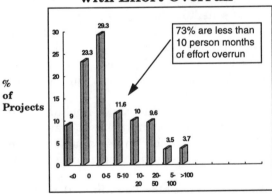

Person Months of Effort Overrun

Figure 8-2. If we accept an overrrun of five person-months as approximately meeting the plan, then about one third (38.4 percent) substantially exceeded the effort and cost plan.

Augustine seems to have conceived his factor while contemplating schedule overruns. The factor seems equally applicable to other estimated quantities, particularly size. In fact, size is the first quantity to be estimated in software development. The length of the schedule and the amount of effort then depend in part on the quantity of work to be done, as roughly measured by size.

Something in the human psyche seems to cause us to misjudge the magnitude of a new task. We recall one first-line manager—a student of practical human nature—who claimed a factor of 2.00. "When one of my people tells me a job will cost so much, I routinely multiply his number by two," he said.

"The fraction one third seems to have enduring scientific significance in determining the schedule error associated with predicting major events in business undertakings," Augustine explained slyly. He labeled "1.33" the "universal fantasy factor."

A division general manager was locally famous for his "Burke factor." He regularly added 7 percent to cost estimates when they reached his level. That was after first-line managers added their various factors, after overhead, after contingencies, after profit. "Simple," Burke said, "that is the average amount by which we missed past estimates." The Burke factor was an early instance of rough and ready calibration. Burke had gone to kindergarten, high school, and college!

Why do these fantasy factors persist?

- Most people are congenitally optimistic (you have to be to get out of bed in the morning).

- People want to please their superiors (the fact that you are still here means you are still eating).

- People don't like to stick their necks out (otherwise we'd be built like giraffes).

- Many people want to avoid confrontation (a big enough size estimate would put you on a collision course with marketing).

- Most people don't think a year or two ahead (when the non-thought-stuff hits the fan).

- Many people want business right now (a bird in the hand is worth two in the bush, our ancestors found out).

- People want themselves and their things to look nice (in the morning the males shave their faces; in the afternoon they shave bids—so the bids will look nicer).

- People want to do dashing things like visit Mount Everest, the Moon, and Mars, or get this bid (let a thousand flowers bloom and water them later).

- People have a worm's eye view of their past projects (one reason you want to have several experienced hands participating in the next estimate).

- People's memories are deceptive (they have glowing memories of the amount of work they did when they were younger).

- People are proud of their splendid memories (and don't like to brush off cobwebs looking up old records).

- People talk and think in single-point round numbers—for example, I was going 60 miles an hour. In ordinary human discourse, we make allowance for the fact that such round numbers usually represent a range. That is part of our human common sense. Somehow, when we embed exact numbers in a six-inch-high bid, we forget that the numbers are still fuzzy.

- Robots may overcome these human frailties some day. In the meantime, we have to estimate size taking them into account. It seems that we may do better if

  ❏ We take steps to get a clear picture of what is to be done (if we don't know what we're going to do, we can't estimate the size of it).

  ❏ We assign several experienced people to draw up the estimate (your strength may offset my frailty).

  ❏ We use a range estimate and statistical methods (making the uncertainty explicit).

  ❏ We get a formal estimating procedure on which all parties agree (confining arguments to the substance of the numbers).

  ❏ We insist that the technical estimate be realistic (play out the political games at the level where executives have the means to understand the game and the profit-and-loss responsibility to be accountable for the results of it).

References

[1] N.R. Augustine, *Augustine's Laws*, Penguin Books, New York, N.Y., 1987, 484 pp.

[2] J.B. Dreger, *Function Point Analysis*, Prentice Hall, Englewood Cliffs, N.J., 1989.

[3] B.W. Boehm, *Software Engineering Economics*, Prentice-Hall Inc., Englewood Cliffs, N.J., 1981, 767 pp.

[4] J.A. Adams, "Military Systems Procurement: The Price for Might," *IEEE Spectrum*, Nov. 1988, pp. 24–33.

[5] C. Woodward, "Trends in Systems Development Among PEP (Productivity Enhancement Programme) Members," PEP Paper 12, Butler Cox PLC, London, Dec. 1989, 65 pp.

[6] J. Rothfeder, "It's Late, Costly, Incompetent—But Try Firing a Computer Systems," *Business Week*, Nov. 7, 1988, pp. 164–165.

Chapter 9

The Fifth Factor: Manpower Buildup

"The two guys trying to write this book apparently missed high school," "Honest Phil" said to John.

"That's silly," John demurred. "They are obviously fairly clever."

"Possibly a bit too clever," Phil replied. "They now have four factors in their software equation: size, time, effort, and process productivity. When you begin estimating, all four are more or less unknown."

"You haven't been paying attention," John interrupted. "Process productivity, they obtained by calibrating past projects. Size, they just estimated in the last chapter, again depending heavily on experience."

"Yes, but they still have two unknowns in their equation: time and effort," Phil rejoined with a triumphant smirk. "The important thing I learned in high school is that you can't find two unknowns when you have only one equation, and they have only one equation."

"Yeah, Mr. Bixler was famous for putting a trick problem like that in his finals," John agreed. "Still, the authors must have gone to high school."

Yes, we went to high school. Moreover, we learned one more thing in high school that Phil seems to have forgotten. Our Mr. Bixler taught us that, when we had two unknowns and only one relationship, we had to dig around in the underlying reality for a second relationship. In a so-called "word problem" in school, we had to pore over the words, trying to sort out another relationship. In software estimating we must analyze project data, looking for another way to relate time and effort.

Manpower buildup

The second relationship between time and effort turns out to be the rate of staff increase during the main buildup. We call it manpower buildup. In formal terms,

Manpower Buildup = Total Effort/(Development Time)^3

If you look at a Rayleigh curve (such as that in Figure 3-4), you can clearly see that the rate of manpower buildup affects both the duration of the curve (development time) and the area under the curve (effort). Suppose the rate builds up sharply. Fast buildup brings more effort under the Rayleigh curve early in the project. The work gets done sooner, leading to shorter development time. Conversely, if the rate builds up slowly, it defers some of the effort until later in the project; and the work takes longer. This pattern is illustrated in Figure 9-1 showing the number of staff on the vertical axis against development time on the horizontal axis. The five lines correspond to the manpower buildup indexes 1 through 5.

Manpower Buildup Index

Management Scale -3 to +10

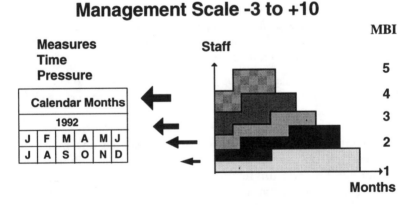

Figure 9-1. Manpower buildup is related to the rate of staff increase during the rising part of the Rayleigh curve. With rapid buildup, such as the line labeled 5, the effort gets crowded into a shorter time.

Often a small manpower buildup rate (a slow buildup) corresponds to a small team; a large rate to a large team. However, this small-team large-team relationship does not always hold. Very small projects are different from larger projects in the way they apply effort.

As in process productivity, an organization normally obtains its manpower buildup by calibrating from past projects. When it knows development time and effort, it can calculate manpower buildup.

Manpower buildup index

Table 9-1 lists the numerical values of manpower buildup. Each value corresponds to a manpower buildup index, also listed in the table.

Management influence

Although managers can influence the manpower buildup rate to a limited degree, the rate is largely dependent on the nature of the work the organization does.

Table 9-1. The most commonly occurring manpower buildup indexes are from 1 to 6. The very small and very large values let us accommodate extreme cases.

Manpower Buildup Index	Manpower Buildup Value	Significance
-3	0.5	
-2	1	
-1	2	
0	4	
1	8	Slow
2	16	Moderately Slow
3	32	Moderate
4	64	Rapid
5	128	Very Rapid
6	256	Extremely Rapid
7	512	
8	1,024	
9	2,048	
10	4,096	

One is that early in the project "a few good men or women" have to partition the work into chunks suitable for one person. How fast this breakdown can be accomplished depends on the "difficulty" of the proposed system and the skill and experience of the few good people. In the short run both the difficulty of the organization's typical work and the skill and experience of the initial people are rather fixed. Management can do little in the near term to improve either.

Another factor is where the task breakdown falls on a scale extending from concurrent to sequential. If tasks have to be done in sequence—if a designer can't start a task until she knows the outcome of the previous task—then fewer people can be usefully employed. The manpower buildup rate is naturally slow. Of course, an impatient manager can throw people at the project, but there won't be tasks ready for them to work on. You expend people hours, but get little product.

If the nature of the work permits many tasks to be done concurrently, more people can be usefully employed. Manpower can build up more rapidly. The upper limit to this buildup is still the number of tasks available for assignment. Management can build up staff more slowly than this limit. Tasks can wait until people are available.

Again, the rate of manpower buildup depends mostly on the nature of the work. The concurrent-sequential scale sets a limit on management's ability to build up rapidly—and usefully.

A third factor is the organization's staffing style. In recent years executives have taken much satisfaction in lean organization. They try mightily to fully occupy everyone. It also means they often have no one available for the next task on a project that is building up. The task must wait until a person who fits the assignment completes what he or she is already doing.

Of course, tasks are inanimate and know neither patience nor impatience. They can wait. But the manpower buildup rate falls, and the new project stretches out. And that may not be what the chief of a lean organization most wants.

If you can stash a few people away on something useful, but interruptible, you may be able to come closer to building up new projects as rapidly as the nature of the work allows. You might have a few people working on background material (requests for proposal that haven't come in yet). You might have some giving or taking training. You might have some developing specialized tools. You might have a few figuring out how to improve your software process. It's a tradeoff.

If you feel especially heavy time pressure on one project, the most effective action you can take early in its schedule is to staff it as rapidly as tasks become available. You must get the people by delaying lower priority work, however. You can bring in people from outside your organization, but that takes time, too. It is hard to find people who fit in right away.

"Big deal," Phil said to John. *"So they've got two equations now. I suppose pretty soon they'll estimate something with them."*

"Patience, impatient lad," John replied.

Chapter 10

Standardize, Discipline, and Use Measurements

Measurement is the soul of software management. [Lord Kelvin didn't say this, but someone should.]*

A manager fighting daily to keep costs below the price the market sets hardly needs a reminder that measurement is a key factor in business. A good manager must keep track of material costs, wages, production schedules, investment costs, and so on, if profits are to be achieved.

In software development unfortunately there has been much soft-headed talk about programming being a form of artistic expression. This talk has led some people—especially those without profit-and-loss responsibility—to forget that programming, when it occurs within a business framework, must be done *on time and within budget, and produce a reliable product.*

---

*We've been collecting pithy sayings on measurement in eager anticipation of this chapter. Then we recollected that "brevity is the soul of wit" (and perhaps of chapters on measurement as well). At any rate, we decided to showcase the best of our collected sayings in this footnote. You can easily skip over them—if you really believe that brevity is that important.

"There is a lot to be done [in software] ...to approach the precision with which alternative ideas can be quantitatively compared, as in engineering." *Les Belady, chairman, Mitsubishi Electric Research Laboratories (in Foreword).* [1]

For a task such as programming to be predictable, it must first be measurable." *Capers Jones, chairman, Software Productivity Research, Inc.* [2]

"Designing software ...must be nudged from an art form to a measurable process before it can be rigorously controlled." *Les Shroyer, chief information officer, Motorola Corp.* [3]

"...The development of adequate measures of both the software itself and the software development process is essential to the production of cost-effective, timely, and reliable software." *S.D. Conte, H.E. Dunsmore, and V.Y. Shen* [1]

The business attitude means the software process must be "measured." In the earlier chapters we learned that an organization needs only four basic metrics to estimate and control software development:

❑ Size of product (usually source lines of code)
❑ Schedule (or duration or development time)
❑ Effort (person-months or person-years, or the equivalent in money)
❑ Defects per time period (usually months).

From the first three metrics, we compute *process productivity*. From the second and third, we calibrate *manpower buildup rate*. From the last one, defects per time period, we compute *mean time to defect*.

From these basic measured and computed metrics we can derive many others for various purposes, such as number of people, code production rate, and cost per month.

However, measurement requires three important things:

❑ You must standardize the basic metrics.
❑ You must discipline your organization.
❑ You must make the metrics usable.

Standardize the metrics

Standardization starts with uniform definitions of the metrics within your particular organization. The definitions should be unambiguous, countable, and precise. We have found the following useful.

Size

Size refers to the quantity of information or functionality in the software system or product. It is measured by the number of *effective source lines of executable code* developed and/or delivered. It includes estimate of equivalent new lines in reused or modified modules that a project has worked on. It does not include comment and blank lines, environmental or scaffolding code, or inherited code that the project has not worked on. In figuring size, count to the line of code; *do not round off.*

Source lines of code (or source statements) are the most common measure of program size, but other metrics—including function points, number of programs, number of subsystems, and number of files, reports, and screens—have also been used.

Schedule

Schedule refers to the elapsed calendar time from the beginning of some phase of development to the end of that phase. The phase of greatest interest is the *main build*, also known as *development time*. Schedule is measured in calendar time, usually months, but also weeks or years. Count to the nearest tenth of a month or equivalent.

In our definition, we exclude feasibility study, functional design, and maintenance (repair or enhancement) from the main build, or development time.

Measuring time is easy, of course. The difficulty lies in establishing unambiguous beginning and ending points.

- ❏ *Beginning of main build*: Start of detailed logic design, sometimes established as the point of completion of the formal review of the functional design or specifications.
- ❏ *End of main build*: The point at which the system or product enters regular, routine use or is formally turned over to the customer. Full operational capability.

If you do not carefully and consistently define and apply these beginning and end points, the basic time metric will be flawed.

Effort

Effort refers to person-months (but can also refer to person-hours or person-years) applied to the project during the main build (or other phase). Effort may be regarded as a normalized version of cost (Person-months × Labor Rate × Overhead Factor). Count to the nearest person-month.

Main-build effort includes work done by all development staff: analysts, designers, programmers, coders, integration and test-team members, quality assurance staff, documentors, supervisors, and managers. It does not include vacation, sick leave, and other nonworking time or time during the project charged to proposals or other unrelated assignments. Generally, it includes all effort charged to the project.

Defects

Count defects found in requirements, specifications, design, and code detected during the development period. An error immediately corrected by the project member who made it before he or she made a record of it is not counted. The defect count is based on a record of review, inspection, or test. After delivery, count defects that occur in program operation. The count is based on a valid problem or trouble report.*

Four metrics are particularly useful in controlling software development and operation:

Total defects. Number of defects counted over the entire period of development and operation. Many organizations classify defects in terms of severity, such as:

❑ *Critical*: prevents further execution; nonrecoverable; must be fixed before program is used again.

❑ *Serious*: subsequent answers grossly wrong or performance substantially degraded; user could continue operating only with allowance for poor results; should be fixed soon.

❑ *Moderate*: execution continues, but behavior is only partially correct; should be fixed in this release.

❑ *Cosmetic*: tolerable or deferrable, such as errors in format of displays or printouts; should be fixed for appearance, but may be delayed until convenient.

Defect rate. Defects per time period, usually per month. This metric is useful for control.

Defects remaining at delivery. Estimated number of defects remaining in the program at full operational capability or delivery. At this time, because you know the number of lines of source code, you can calculate the defects remaining per thousand lines of source code. This metric is useful for comparison among projects.

Mean Time To Defect, or MTTD. During development, the mean calendar time from the finding of one defect to the next. This time is, of course, the reciprocal of the defect rate. During system operation, the mean time, usually in operational or execution time, from one failure to the next; hence often called Mean Time To Failure, or MTTF.

Figure 10-1 illustrates the metrics pertaining to defects. The figure also indicates the reliability levels, the points in the life cycle at which 95, 99, and 99.9 percent of the defects have been removed. Developers often deliver at the 95 percent reliability level but could continue defect removal to reach a higher reliability level.

---

*Actually, for simplicity, we term all errors, faults, and failures during development and operation, defects, Error, for instance, carries the connotation of human fallibility in requirements analysis, specification preparation, design, coding, or test planning. An error leads to a fault in a program that causes a failure in test or operations. [4]

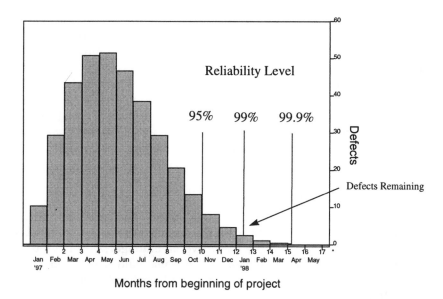

Figure 10-1. On a Rayleigh curve of the defect rate each short interval represents defects per time interval (such as month). The area under the curve represents the total number of defects. After delivery at the 95 percent level, the area under the curve represents the defects remaining.

One of the wonderful things about software metrics is that you can start wherever you are. If you have only one software organization within your purview, you can standardize the measurements it makes. Then your estimates will be more consistent because the measurements from which they ultimately derive are more consistent. Comparisons with the results obtained by other organizations may be flawed to the extent that your measurements are not consistent with theirs.

If you have executive responsibilities larger than one software organization, you can standardize more broadly. Then you can confidently compare progress in achieving estimates across all the projects within your scope.

Still better, if you can standardize in line with industry-wide definitions, you can compare your operations with comparable application categories or industries in our database.

Failure to define counting rules for source lines of code has led to variations of "as much as 5 to 1 between the most diffuse counting technique and the most compact," according to Capers Jones [2]. Combining the range of variation of the main software metrics may introduce "potential errors in excess of two orders of magnitude." Obviously, these and other flaws in measurement practice make a mockery of estimating procedures.

In our experience, rounding off measurements has by itself introduced variations from the true value of from 10 to 20 percent. If estimates of future work are to be good, past measurements must be precise. Instead of rounding off source lines of code to the nearest thousand (or five thousand or ten thousand), count to the actual number. Instead of rounding off schedule to the nearest month (or nearest quarter, half year, or year), measure to the nearest tenth of a month—19.4, not 19. Instead of estimating person-months, count the actual person-months, or get them from the accounting records.

Discipline your organization

You may have noticed that organizations do not automatically use standards. If you are the responsible executive or manager, you have to set the example. You need to inquire about, discuss, and use the measurements. You have to maintain a continuing interest over the procedures used to collect statistics.

You should not use these measurements to evaluate people. When used for that purpose, people tend to twist the numbers over time, as they figure out how to make themselves look good. Because of the crucial importance of controlling software development, you are not interested in a few perhaps overly clever individuals "looking good." You are interested in organizational achievement.

Make metrics usable

In the days before standardization individual managers, supervisors, and estimators tended to keep some numbers they found useful in little black books or desk drawers. These were the numbers they used to make those grossly optimistic estimates. More complete numbers on past projects, if they existed at all, were in unorganized form in a warehouse somewhere. No one had ever been out there!

In the new day you want to use the new and more precise numbers. So you have to make software measurements readily accessible. Because you use them to compute process productivity and staff buildup on past projects, and to compare process improvement over time, they have to be available for many years. In this day of computers, accessibility means putting them into a database.* Then everyone (or everyone who needs estimating data) can get at them quickly.

---

* We invite those interested in pursuing the project database idea in more detail to read our technically oriented book *Measures for Excellence: Reliable Software on*

An old-timer reminisces: "I can't count the number of times I've watched software teams set up a metrics database, put measurements in it, and then never use it. Frequently I hear: 'We collect metrics because our standard requires it.' Or, 'we don't collect metrics on this software because it doesn't get delivered and thus doesn't fall under our standards. We only use this software to test the software we deliver.' "

"Projects will start reducing staff because they are over budget, because the plan said they should start downstaffing, or for some other reason totally unrelated to whether they now have less work to do. The metrics database could have helped answer this question."

"This is all like trying to drive a car with a dirty windshield and no windshield wipers—you don't know where you are headed. And no rear view mirror—you don't know where you have been either."

How to get started in metrics is the province of a 1987 book by Robert B. Grady and Deborah Caswell, recounting the experience of Hewlett-Packard [5]. Five years later Grady recounted further lessons learned [6].

Referring to IBM's total quality program, former CEO John Akers wanted IBM not merely to satisfy its customers but also to 'delight' them. There is one difficulty, IBM executive Stephen Schwartz admitted. 'We're still trying to figure out how you measure delight.' —from an article in Fortune [7]

References

[1]   D. Conte, H.E. Dunsmore, and V.Y. Shen, *Software Engineering Metrics and Models*, The Benjamin/Cummings Publishing Co., Inc., Menlo Park, Calif., 1986, 396 pp.

[2]   C. Jones, *Programming Productivity*, McGraw-Hill Book Co., New York, N.Y., 1986, 280 pp.

[3]   G. Rifkin, "No More Defects," *Computerworld*, July 15, 1991, pp. 59–62.

[4]   S.L. Pfleeger, "Measuring Software Reliability," *IEEE Spectrum*, Aug. 1992, pp. 56–60.

[5]   R.B. Grady and D. Caswell, *Software Metrics: Establishing a Company-Wide Program*, Prentice Hall, Englewood Cliffs, N.J., 1987, 288 pp.

Time, within Budget (Prentice-Hall, 1991). Also, Quantitative Software Management provides such a database system, Productivity Analysis Database System, complete with a manual and training course. PADS contains information on the projects in the QSM database with which you can compare your projects.

[6] R.B. Grady, *Practical Software Metrics for Project Management and Process Improvement*, Prentice Hall, Englewood Cliffs, N.J. 1992, 270 pp.

[7] F. Rose, "Now Quality Means Service Too," *Fortune*, Apr. 22, 1991, pp. 97–111.

"The development process must be disciplined."

Part III

Project Management

You can plan software effort and schedule realistically. Or you can gather hard-won experience, as IBM did at the beginning of the computer age. The IBM System/360 was the largest software development of its time (the 1960s). It is the project on which Fred Brooks learned the lessons that he so charmingly recounted in *The Mythical Man-Month* [1]. In 1966 IBM's software budget was $40 million and 2,000 people were working on the 360's basic software, according to Thomas J. Watson, Jr., then IBM's chief executive officer [2].

"By the time the 360 software was finally delivered, years late, we'd sunk a half-billion dollars into it alone, making it the single largest cost in the System/360 program, and the single largest expenditure in company history," Watson said.

"We learned the hard way one of the great secrets of computer engineering: throwing people at a software project is not the way to speed it up," Watson explained further. "A piece of software is a unified thing; if you try to break up the job of writing it among too many people, it takes more time to coordinate them than the division of labor saves."

You probably have friends who have had experiences such as IBM did—and more recently. With the knowledge of estimating, planning, and controlling software projects the following chapters outline, you can get much closer to realism.

References

[1] F.P. Brooks Jr., *The Mythical Man-Month: Essays on Software Engineering*, Addison-Wesley Publishing Co., Reading, Mass., 1974, 195 pp.

[2] T.J. Watson Jr. and Peter Petre, *Father Son & Co.: My Life at IBM and Beyond*, Bantam Books, New York, N.Y., 1990, 468 pp.

Chapter 11

Planning Effort and Schedule

"More software projects have gone awry from management's taking action based on incorrect system models than for all other causes combined." —Gerald M. Weinberg [1]

"It looks like the office automation system is going to take longer than we first thought," the software manager told the general manager. "The minimum development time calculates out to be 21.6 months. On that schedule the development effort would be 392.3 personmonths with a peak of 28 people. Cost would be $3,297,000. These are the statistically expected values—we have a 50 percent probability of hitting them."

"That's longer than I had hoped for," the general manager mused. "We sure need the savings it promises to bring."

"The statistics give us very little chance of beating that time," the software manager said.

"I know. I've been through computations like this before," the general manager said. "Besides, we have to order a million dollars worth of hardware, staff a new computer unit, and train people in the new methods. We have to run the old manual system and the new system in parallel for some months. It is important that the software be ready on the day we plan to start up. Otherwise all that hardware and all those people sit around running up costs."

"We ought to pick a combination of effort and time, then, that increases the probability that the software will be fully operational on time," the software manager offered. "Allowing a development time longer than the minimum does that."

"Do you have a number in mind?" the general manager asked.

*"Yes, I think it would be reasonable to extend the planned develop-
ment time to 24.0 months," the software manager replied, shuffling
through his computer printouts. "Lengthening the planned schedule
not only reduces overall effort and the corresponding cost, but it also
reduces the expected number of errors. With fewer remaining errors
after the system goes operational, the conversion will go more
smoothly."*

*"I know, and that is all to the good, but your probability of completing
in 24.0 months is still only 50 percent," the general manager pointed
out. "That leaves us one chance in two that the new hardware and
new staff will be sitting on their butts while you try to finish the
software."*

*"You're right, of course. If we plan a 24.0- month schedule, we would
have a 70 percent chance that the software will be ready to go in 24.7
months. We would have a 90 percent chance at 25.8 months," the
software manager said. "What level of certainty do you think you
need?"*

*"I'm not going to answer that off the top of my head," the general
manager chuckled. "That's the kind of answer that ought to come
from an analysis. What is the probable cost of various software development plans versus the cost of letting the new facilities stand idle?
Get somebody to work it out."*

"Yeah, Ralph can do it."

*"I'll toss one intangible into his analysis process," the general man-
ager interjected. "He might not know about it. I don't think anyone
else in the industry is working on a system like this one yet. We can
afford to lean in the direction of economizing on effort and being
fairly certain of the delivery date. That would set the startup date
back some and delay getting the savings from the new system, but it
wouldn't hurt our competitive position."*

The two managers had discussed this project before and will discuss
it again. This fragment of their ongoing discussions is not all there is to
say, but it is notable in several respects. Both managers are working off
the same estimating method, so they easily understand each other.
There were no arguments or hard feelings.

Both talk in precise numbers—to the nearest tenth of a month, for
example. At first glance, using precise numbers seems strange, because
they also talk in terms of probabilities. The estimates are not really precise. They are *expected* numbers within a probability range. Both managers are in the habit of using precise numbers because they collect project data that way. Accurate past data provides the basis for projecting

accurate numbers on new projects. They realize that the more precise the past data, the better new estimates will be.

Moreover, the use of probabilities quantifies the uncertainties inescapably involved in developing software. Putting numbers on these uncertainties gets them out in the open and removes one more cause of argument.

How do you acquire the easy familiarity with estimating practice that these two managers exhibit? In this chapter we begin to find out.

Visualizing effort and time

In Part II we developed two equations that relate effort and development time, repeated here for convenience:

$$(E/B)^{1/3} * t_d^{4/3} = \text{Size/PP}$$

and

$$K/t_d^3 = \text{MB}$$

where

$$
\begin{aligned}
E &= & effort \\
B &= & a\ constant \\
t_d &= & development\ time \\
\\
\text{Size} &= & source\ lines\ of\ code \\
\text{PP} &= & process\ productivity \\
K &= & E/.39 \\
\text{MB} &= & manpower\ buildup
\end{aligned}
$$

The effort and development time terms on the left side of the equations are the unknowns at this point. We have estimated size and we know the process productivity and manpower buildup from calibrating past projects. With two equations and two unknowns, we can now solve for values of effort and development time. But first let's visualize some of the combinations of effort and development time a project might employ.

People might be applied to a project in the pattern of Figure 11-1A. Manpower buildup is rapid, getting a lot of effort under the Rayleigh curve during the short development time. In Figure 11-1B manpower is built up more slowly, resulting in less effort under the curve, but a longer development time. We could draw a number of intermediate curves between these two examples and select one of them as the plan of effort and development time to be employed.

Figure 11-1A implies that there is an upper limit to manpower buildup. At the extreme we could imagine that manpower buildup is nearly vertical, making the development time very short. At the same time the number of people would be very large. That is obviously unrealistic.

Similarly, for Figure 11-1B we can imagine manpower buildup as being nearly horizontal. That would make the development time very long and the effort very small. That, too, is unrealistic. We conclude that there is some range between these two extremes at which the effort-development time pairs are practical.

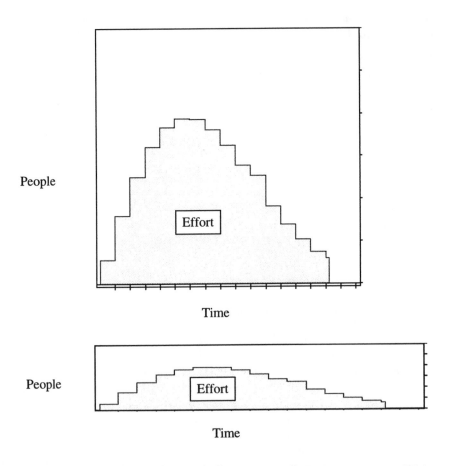

Figure 11-1. Two Rayleigh curves show two patterns of manpower buildup. **(A)** In this pattern, the manpower buildup is very rapid, getting considerable effort applied early in the project. **(B)** A more gradual manpower buildup spreads effort over a longer period. Most practical effort-time patterns lie somewhere in between.

Locating the practical region

The development time and effort of a proposed system falls somewhere on a field of effort vs. development time, as Figure 11-2 illustrates. We can imagine the effort-time pairs of all possible projects being located somewhere on this field. There would be dots covering most of the diagram. As a matter of common sense, however, we would expect few dots near the vertical axis where time approaches zero and few dots near the horizontal axis where effort approaches zero. All projects require both time and effort.

For a system of a given expected size and process productivity, there is a range of effort-time pairs at which it might be built. Figure 11-3 shows this range in the form of a line, which we call the size-divided-by-process-productivity line, or *Size/PP line*. We cut off the line at each end because we know, as a practical matter, that effort-time pairs cannot reach either zero effort or zero time. As a matter of fact, the scatter plots in Chapter 4 show that projects of more than a few thousand lines of code require finite amounts of time and effort.

Effort - Time Relationship

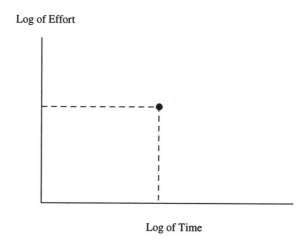

Figure 11-2. All projects may be located somewhere on the field of effort and development time, but few are near the vertical and horizontal lines because all projects take at least some effort and some time.

Effort - Time Relationship

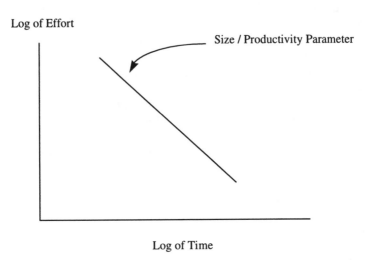

Figure 11-3. Possible effort-time pairs at which a system of a given size and process productivity might be planned fall on a Size/PP line. The software equation underlying this figure contains exponential values of effort and time. When plotted on log-log scales, the Size/PP line turns out to be a straight line.

Note that as projects allow more development time the corresponding effort decreases.

In Figure 11-3 we can visualize a considerable range of effort-development time pairs along the Size/PP line. To the left of the left end of the line—wherever precisely that is—there are no effort-time pairs. We call that the "impossible region." No one has ever come back from this region with a successful project. In the next chapter we will examine this region and locate its edge.

Reference

[1] G.M. Weinberg, *Quality Software Management: Vol. 1, Systems Thinking,* Dorset House Publishing, New York, N.Y., 1992, 318 pp.

Chapter 12

The Impossible Region: Keep Out!

When I feel the urge to set an impossible schedule, I lie down until it passes away. [Mark Twain didn't say this, but he probably would have if he were still around.]

It is not impossible for an executive to arbitrarily set a development time that falls in the impossible region. It is even possible for software managers to plan a project in this region. It is impossible, however, to accomplish it.

Perhaps something like this happened to the Internal Revenue Service some time ago: (If you have harbored a repressed resentment of the IRS, here is your chance to snicker.)

In May 1985 the IRS estimated the cost of the Automated Examination System to be $1 billion (including 18,000 laptop microcomputers), according to a study by the US General Accounting Office. The system was to enable revenue agents to examine Form 1040 returns more efficiently. The IRS estimated benefits over the nine-year life of the system to be $16.2 billion. Scheduled completion was 1989.

In January 1987 the project was rescheduled for completion in 1991. The cost estimate grew to $1.2 billion. Again in March 1988 a new analysis extended the completion date to 1995 and the cost to $1.8 billion.

Unfortunately 77 percent of the agents queried expressed dissatisfaction with the initial release. Only one third said they used it. Field managers doubted the software was saving agents' time, throwing the $16.2 billion benefit into question. Responding to these uncertainties the Office of Management and Budget reduced 1990 funding from $110 million to $20 million.

As project members come to appreciate that they are embarked on "Mission Impossible" without a camera crew, they become frustrated. Then they become irritable. Some become angry. They begin to view the executive responsible for the schedule as ignorant of software realities and careless of their own welfare. The more knowledgeable people angle for transfers to other projects.

Or, in Tom DeMarco's phrase, some vote with their feet. "That's what capitalism does best," he said. "It throws out those companies. You can do good by helping your company succeed. But you can also help by helping it fail" [1].

When an executive sets an "impossible goal" and software developers vote with their feet, both are carrying out their modest roles in the inexorable process of "Creative Destruction," as Joseph A. Schumpeter called it. It is the process that "incessantly revolutionizes the economic structure *from within,* incessantly destroying the old one, incessantly creating a new one," Schumpeter explains [2].

The "impossible region" sounds somewhat paltry in the face of the civilization-wide gale of creative destruction that Schumpeter summons up, but a gale consists of many particles. Software planning based on little more than whim is, in any one instance, a small particle indeed. Repeated thousands of times, it changes the face of industry.

What we mean by the "impossible region" is that no comparable software organization (with the same process productivity and man-power buildup) has completed a similar system on the proposed schedule. To put it another way, *there is a minimum development time* in which a given organization can build a system of a given size and application type.

The "impossible region" is not a figment of our fevered imagination. It is solidly based on empirical evidence. Our 3,885-system database contains no projects completed in this region. Barry Boehm referred to the "impossible region" in 1981 [3]. The following year Tom DeMarco regretted that he had spent so much of his earlier life in the region [4].

Ken Orr decided to "do something" about the region! "We've replaced the need for study and thinking with the latest tools," his "One Minute Methodman" said. The new methodology pushed development time to the far, far edge of the impossible region: one minute. All went well for a time until the auditors informed the chairman of the board that "instead of making a record profit last year, as our One Minute System reported, the company actually went bankrupt."

"Well, there are, of course, occasions when projections based on random numbers can be a tiny bit misleading," the One Minute Methodman admitted [5].

Locating the impossible region

If we knew enough about a project to draw a critical path (or PERT) diagram, the length of the critical path would represent the minimum schedule. Thus, in principle, there is such a time as the minimum development time. We could draw a line across the Size/PP line at this point and say, effort-time pairs to the left of this line are in the "impossible region."

In the early stages of a proposal, we don't know enough about it to draw a critical-path diagram. We do know, however, that one factor affecting how long a project takes is the extent to which tasks have to be done in sequence or concurrently. Manpower buildup is, among other things, an expression of this sequential-concurrent contention. People cannot be assigned to a project *usefully* any faster than separate, nonsequential tasks can be sorted out for them to work on. If a project is largely sequential, the work must be spread out accordingly.

Thus, there is some maximum manpower buildup set by the nature of the work itself. We find this maximum rate by calibration from past projects. An organization's history tells it what this maximum buildup is. In general, an organization cannot build up faster than its past practice, at least in the short run.

Buildup is also limited by the skill and experience of the people, the extent to which people are available when needed, the use of software tools, the availability of workstations, and so on—factors that tend to remain the same from project to project. Hence, calibrating against recent history is a valid approach to determining the minimum development time. Of course, in the long run an organization can improve its ability on these factors and learn to build up staff more rapidly.

Therefore, we draw a line representing manpower buildup on the diagram, as Figure 12-1 shows.[*] This line, labeled MBP (manpower buildup parameter), cuts off the Size/PP line. The point of intersection is the minimum development time. The region to the left of the MBP line is the "impossible region." It is the region to stay out of!

The exact location of the MBP line is itself uncertain, a topic we explore in the next chapter.

---

[*]The equation ($MB = Total\ Effort/Time^3$) for manpower buildup is also expressed in effort and time. Therefore, we can draw it as a straight line on the log-log field of effort and time. A single value of MBP, represented by the line, can have various values of effort and time. All this—equations, log log planes—may sound forbidding. Remember, we derived the equations from the database. They stand for real experience. When we come full circle, to the impossible region of Figure 12-1, all we are really saying is that the database contains no points in that region.

Minimum Schedule Concept

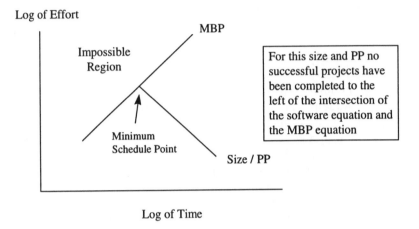

Figure 12-1. Where the line of manpower buildup (MBP) cuts off the Size/PP line is the point of minimum development time. To the left of the MBP line is the impossible region.

Dealing with business pressures

Knowing that there is an impossible region and that you can compute where it is might keep you clear of Schumpeter's destructive gale. Right now, however, you are facing overpowering business pressure to get the job done in a shorter time? How do you respond?

Actually, the development time at the intersection of the MBP and Size/PP lines is "minimum" only under prescribed conditions: a given size, process productivity, and manpower buildup. If you can change these conditions, you can shorten the minimum development time.

If you reduce the size of the proposed system, the Size/PP line moves downward to the left, as Figure 12-2 shows. The minimum development time shortens, as shown by M'. Of course, lessening the size and functionality of the desired system introduces a different set of problems.

If you can improve your process productivity, the Size/PP line follows a similar migration downward to the left. Unfortunately for the project at hand, improving process productivity enough to shorten the minimum development time generally takes a period very much longer than a project schedule.

Trade-off Concepts
Size or PP

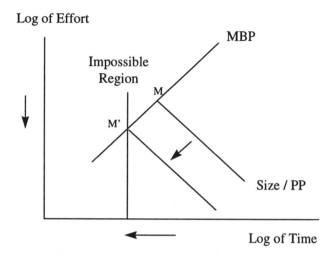

Log of Effort

MBP

Impossible
Region

M

M'

Size / PP

Log of Time

Figure 12-2. Reducing size (or increasing process productivity) moves the Size/PP line downward to the left, shortening the minimum development time to M'.

Complexity can also move the Size/PP line. A successful effort to limit product functionality, such as reducing the number of bells and whistles, at the same time reduces size. Limiting the number of interactions among product parts makes it easier to design and build, improving process productivity.

Finally, if you can increase your manpower buildup, the MBP line moves to the left, as Figure 12-3 shows. The minimum development time shortens to M'. Again, because manpower buildup depends greatly on the nature of the work you do, it is not readily increased in the short run. You cannot build up more rapidly just by getting out a whip. Software experts must figure out difficult problems such as how to partition the work more quickly and do more things in parallel. Dependencies become very important issues.

Usually a particular piece of the software process, such as partitioning the work or reducing complexity, does not occur in isolation from other pieces of the process. In Part V we consider process improvement as a whole.

Trade-off Concepts
Staffing

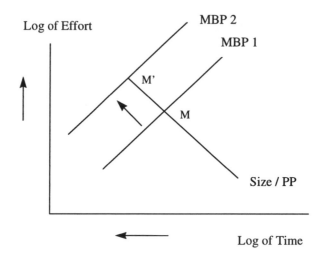

Figure 12-3. Trading off staffing. Increasing manpower buildup moves the MBP line to the left, reducing the minimum development time to M'.

So the courses of action the diagrams indicate take a good deal of time. The actions you might have taken in the past, like throwing more people at the project, don't seem to work very well in software development. It is the kind of situation for which Fred Brooks forged a law: "Adding manpower to a late software project makes it later" [6].

References

[1] G. Gruman, "At ICSE 13, Diversity Is The Theme," *IEEE Software*, July 1991, pp. 97–98, 100.

[2] J.A. Schumpeter, *Capitalism, Socialism and Democracy*, Harper & Row, New York, N.Y., 1942, 431 pp.

[3] B.W. Boehm, *Software Engineering Economics*, Prentice-Hall Inc., Englewood Cliffs, N.J. 1981, 767 pp.

[4] T. DeMarco, *Controlling Software Projects*, Yourdon Inc., New York, N.Y. 1982. 284 pp.

[5]   K. Orr, *The One Minute Methodology,* Dorset House Publishing Co., New York, N.Y., 1984, 59 pp.

[6]   F.P. Brooks Jr., *The Mythical Man-Month: Essays on Software Engineering,* Addison-Wesley Publishing Co., Reading, Mass., 1974, 195 pp.

"This is one bridge we're going to build in less than the minimum development time."

Chapter 13

Risk is Ever Present

Risk is inherent in the commitment of present resources to future expectations.—Peter F. Drucker [1]

"Business is the art of putting together something we know how to do in a planned time with an accurately forecast expenditure of resources," the businessman said. "The good businessman keeps the margin of error small enough to permit a profit."

"Art is the business of creating something unique in an indefinite amount of time and effort," the artist said. "The profit is in the pleasure it gives."

Sometimes the programmers in the back room have a little trouble keeping this distinction in mind. Even on a business basis, there are uncertainties in developing software. They may not matter to software as art. They do most certainly matter to software as business. Therefore, in estimating resource expenditures it is important to take the uncertainties into consideration.

In 1988 the US Army canceled a program to automate its civilian personnel records. The reason given was schedule and cost overruns. The project may have been overly ambitious for the time and funds available.

"I assign myself blame," Major General Alan Salisbury told the Washington Post, "for not standing up and instead saluting and saying, 'Yes sir, we'll build that system.'"

The uncertainty of minimum development time

The minimum development time is not a firm, one-point number, convenient as that would be. It is an "expected" value. "Expected" is a statistical term meaning that there is only a 50 percent probability that you can complete the project in the expected minimum time.

On the one hand, it is possible to rerun the computation at a development time longer than the minimum. That will give you better odds.

On the other hand, perhaps those irresistible pressures are forcing you into one of those painful gambles. You feel obligated to quote a time shorter than your history warrants. You can run this too-short time value through the computations and find, for instance, that you have only a 10 percent probability of completing on that schedule.

At least you will not be surprised when the project comes out that way. You can be readying your excuses or keeping careful track of changes that will support your argument for more time or buttering up your customer or contributing to your Congressman or taking your headhunter to lunch—whatever. You've got time to plot your strategy.

You might also give some thought as to whether a reduced functionality product might meet your customer's near-term needs. You could give those parts of the system priority. Having important parts of the system ready at the due date might mollify the customer sufficiently for you to live on.

Dealing with uncertainty

Computing estimates of development time and effort takes place along the lines of the process in Figure 13-1. All three inputs to the New Estimate Computation block are themselves uncertain to some degree.

As we saw in Chapter 8, the size estimate represents a range of possible values. The number finally stated is the "expected value." That term means that the probability is 50 percent that the value is greater than this number, and also 50 percent that it is less. This in turn means that when dealing with an "expected" value, the probability is 50 percent that the project will turn out unhappily. For example, shortly after an earthquake of 7.4 magnitude in Southern California, a scientist at the Seismological Laboratory at the California Institute of Technology told the television audience: "There is a 50 percent chance of an aftershock of at least 6.0 magnitude today." The next day, when no aftershock of any such magnitude had occurred, the anchormen and women seemed a bit miffed. It was left to a newspaper columnist the following day to point out: "There was also a 50 percent chance of *no* big aftershock."

Software people need to remember that bit of insight.

Dealing with Uncertainty

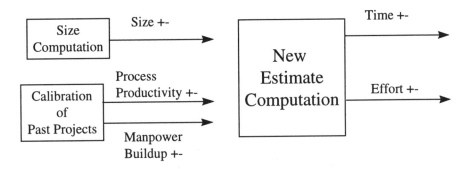

Figure 13-1. Uncertain inputs—size, process productivity, and manpower buildup—result in uncertain estimates of development time and effort.

An expected value is usually graced with another statistical ornament: the standard deviation. It indicates the extent of the range around the expected value. It is an indicator of the uncertainty of the expected value.

Figure 13-2 is a graph of these relationships. The expected value falls on the 50 percent line at 70,000 SLOC. There is less than a 1 percent chance that the ultimate actual value will approach 100,000 SLOC. The +1 standard deviation line is the 84 percent probability level; the -1 line the 16 percent probability level.

The line of possible size values has a steep slope, indicating that the size estimate has a large range. In fact, in this example, it represents a ballpark estimate, made in the feasibility phase when little information was firm. A later estimate, based on more knowledge about the project, would have a shallower slope. In general, the steeper the slope, the more the uncertainty.

Similarly, even though we usually obtain process productivity and manpower buildup by calibration, their magnitude is still not absolutely certain. Past managers may have inconsistently disciplined data collection. Earlier projects may have recorded data inaccurately. Moreover, when records are not available, we must estimate process productivity and manpower buildup from judgments of pertinent attributes. In this event, process productivity and manpower buildup are even more uncertain.

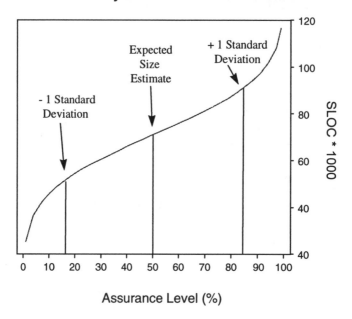

Probability that size will not exceed

Figure 13-2. The sloping line of potential size values relates a range of system size on the vertical axis to the probability of not exceeding each size on the horizontal axis.

When we enter three numbers, uncertain to various degrees, into the New Estimate Computation, the desired outputs, time and effort, are also uncertain, again indicated by + or - on the figure.* The time and effort estimates also consist of two numbers: the expected value and its standard deviation.

Getting risk in the picture

A practical executive can distinguish between proposed systems that are based on known factors and proposed systems with unknown elements. With known factors he can estimate management numbers fairly accurately. With unknown elements he cannot estimate management num-

---

* The New Estimate Computation employs a method called Monte Carlo simulation. This method varies the input numbers at random within a range determined by their standard deviations. We run the computation many times, 1,000 or 10,000. The answers given by these multiple calculations spread over a range, too. We average these answers to provide the expected value. From the spread of the answers, we compute the standard deviation of the expected value.

bers with any precision. The first type leads to a successful business. The second type, if undertaken on the basis of mistaken estimates, leads to eventual business failure.

Experienced software managers agree that three items present the greatest risk. The top-ranked risk item is shortfalls in the people available. Next comes unrealistic schedules and budgets. Third is various deficiencies in requirements [2]. In a similar vein, a US Air Force pamphlet listed requirements complexity, personnel, reusable software, and tools and environment as the principal drivers of cost risk [2].

These risks can be reflected in the three inputs to the New Estimate Computation (Figure 13-1): size, process productivity, and manpower buildup. For example, if the requirements are not firm, you cannot predict the size of the eventual system with any precision. The experienced people selected to make the size estimate should, under these circumstances, set a large range between minimum and maximum size. This large size range, after passing through the New Estimate Computation, will be reflected in large standard deviations of the time and effort estimates.

The calibration method of arriving at process productivity and manpower buildup greatly reduces the risk of unexpected personnel and organization shortfalls. You have your people, you have your organization, you have your existing tools and equipment, you have your present ways of doing things—all are reflected in your indexes of process productivity and manpower buildup. You have merely to continue along the same track to hold unforeseen risk to a minimum.

If you do not have recorded experience with which to find these indexes by calibration, you can estimate them by rating personnel and organizational attributes. But ratings involve judgment and judgments are not precise. In effect, an index arrived at by judgment represents a broader range of possible values than one arrived at by calibration. In consequence, the New Estimate Computation assigns a larger risk to these judgment-based inputs, resulting in a correspondingly larger standard deviation for the time and effort estimates.

With large standard deviations, a bid at the center of the time and effort ranges is very risky. By definition, the risk of not completing successfully at "expected" bid numbers is 50 percent. In addition, the cost of failure to complete is greater with a large standard deviation than it is with a small standard deviation.

The statistical methods contained in the New Estimate Computation block of Figure 13-1, however, enable us to compute the increased time and effort (and cost) needed to improve the odds. Time and effort estimates can be computed for odds from 51 to 99 percent. An executive who evaluates a particular bid situation as calling for caution can ask for a bid, for example, with a 90 percent chance of completing at that level. Of course, there is still a 10 percent chance of overrunning that bid.

Conversely, if circumstances require "buying in" to the job, the methods can compute the low odds, such as 20 or 30 percent, of completing the work within low bid numbers.

An organization cannot be absolutely certain of making money on every project. Bidding at the 20 percent level, for instance, it should expect to lose money over a series of projects. At the 50 percent level, it will likely break even over a series. With conservative bidding, it can average a profit over a series.

These methods do not dictate your decision. They merely tell you what the odds are at various bid levels. On the basis of all you know about the general situation, you are the one to bring judgment to bear on the various bid probabilities.

"I don't want yes-men around me. Tell me what you think even if it costs you your job." attributed to Louis B. Mayer, long-time head of production at Metro-Goldwyn-Mayer.

Using point estimates

It is customary in developing commitments between financially or legally separate entities, such as independent divisions or companies, to set prices and schedules in terms of single-point estimates. A man from Mars, noting that everyone knows a point estimate is unrealistic, would find the practice strange.

In fact, "use of a range of estimates allows for software cost risk analysis," the US Air Force said in a pamphlet on Software Risk Abatement [2]. "Analysis should combine uncertainties for the various cost elements, determine software cost range (minimum and maximum costs), identify the 'most likely' or 'best' estimate, and assess the risk of overrunning budget."

In 1974 when he was Assistant Secretary of the Army for Research and Development, Norman R. Augustine spearheaded a program called TRACE. It accepted the fact of this uncertainty and called on program officers to defer some funds "to deal with foreseeable project uncertainties—the known unknowns." Unfortunately, higher levels of the Defense establishment, the Office of Budget and Management, and Congress tended to seize on these contingency funds as a place for painless cuts. Needless to say, the pain did come later.

In 1981 then-Deputy Secretary of Defense Frank C. Carlucci recommended that DoD increase its "efforts to quantify risk and expand the use of budgeted funds to deal with uncertainty."

Nevertheless, point estimates are still the rule. A system for estimating the amount of risk in software projects cannot inaugurate the practice of range estimates, of course. Still, the method can tell you the degree of risk associated with the point estimate you finally decide, on the basis of your intangibles, to go with.

As General George S. Patton once said,

Take calculated risks. That is quite different from being brash.

References

[1]   P.F. Drucker, *Management: Tasks, Responsibilities, Practices,* Harper & Row, New York, N.Y., 1973, 839 pp.

[2]   B.W. Boehm, *Tutorial: Software Risk Management*, IEEE Computer Society Press, Los Alamitos, Calif., 1989, 496 pp.

"Ladies and gentlemen, our days in the tar pit are over."

Chapter 14

Trading Time for Effort

Please be aware that the software equation was derived empirically: It's not anybody's theory about how effort ought to vary as schedule is compressed; it's the observed pattern of how it has varied. —Tom DeMarco [1]

In Chapter 12 we focused on the impossible region marked off by the minimum development time. We noted that the Size/Process Productivity line, sloping downward from left to right, reveals a useful insight. As project planners lengthen the development time, the corresponding effort declines. At the minimum development time, effort is at a maximum. And because cost is proportional to effort, cost is also at a maximum. Defects will be maximum at that point, too, but we defer detailed consideration of defect behavior until Part IV.

In this chapter we show that you can trade off the management numbers. You can trade off more time for less effort and less cost.

The tradeoff region

The three dots on the Size/PP line in Figure 14-1 mark three tradeoff points. On a log-log scale the differences in time and effort from one point to the next may seem unimpressive. However, remember that a log-log display compresses the space between the points. In actuality, a small extension in time results in a substantial reduction in effort, cost, and defects.

Table 14-1 lists the tradeoff values for a medium business system (50,000 SLOC) developed by an organization with average process productivity (PI of 16 and a manpower buildup index of 5). The actual numbers are more impressive than the log-log diagram. At 130 percent of the 7.62-month minimum development time, or 10 months, for example, effort declines by a factor of three. Mean time to defect improves by a factor of four.

Practical Limits to Time-Effort Trade-off

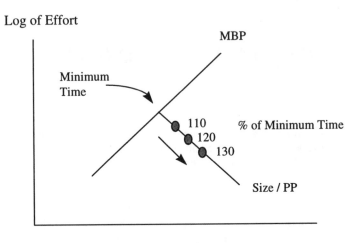

Log of Time

Figure 14-1. There are practical limits to the time-effort tradeoff. It is usually impractical to plan a development time much greater than 130 percent of the minimum
development time. A planned extension of the development time will gain a substantial reduction in effort.

Table 14-1. Extending the planned development time by 1.4 months beyond the
7.62-month minimum development time reduces effort and peak number of people
to about half what they would be at the minimum development time. Mean time to
defect more than doubles.

Time (months)	Percentage Extension	Effort (person-months)	Peak People	MTTD (days)
7.62	Minimum	143	27	0.77
8	105	118	21	0.99
9	118	74	12	1.78
10	130	48	7	3.01

Managers must make the tradeoff decision at the time they plan a project. Then supervisors assign people and carry out the work in accordance with this plan. It is not fair to apply the tradeoff principle when you are three weeks from the end of the very short development time you originally set and you have become painfully aware that you will not complete the system on time.

Of course, you can always replan a project, but you must fit the new plan to the circumstances at that time. The tradeoffs you could have made at the original planning time are no longer available.

The impractical region

Of course you can't go on indefinitely extending the development time and reaping the benefit of reduced effort. The plan would eventually reach a time unrealistically long in terms of market requirements, and it would eventually reach a ridiculously low level of effort. In most cases the practical limit occurs at 130 percent of the minimum development time. Beyond that is the impractical region.

The cutoff at about 130 percent is a matter of judgment. You could allow more time, of course, but gains in effort, people, and MTTD would begin to taper off. The reasons for considering 130 percent as the practical limit are largely pragmatic.

One reason is that the number of people on the project team often begins to approach the minimum feasible size. A project team must represent all the key skills. It must have enough people to carry on in the face of resignations, vacations, or illnesses.

If we carry out the example of Table 14-1 one more month, for example, the peak staff declines to four people. This additional month extends the time to 11 months, or 144 percent of the minimum development time. This level of project staffing probably becomes unacceptably risky, unless you have several good people on nearby projects you could rush into the breach. That becomes a matter of fine judgment, not mathematical formulas.

Another reason lies in the time pressure that besets business activities. The cost of not having a software system may be great enough to justify a schedule at minimum development time or not far beyond it. If the new system is a critical component of a company's competitive strategy—if it is a small cost compared with the computers, machinery, buildings, and people making up the entire system—then you must balance the total cost against the software development cost.

Running a few of the black dots in the practical tradeoff region tells you what the cost and MTTD gains would be. You have some software project numbers to weigh against the system-wide numbers.

Sometimes the time pressure is not as acute as all the hollering and screaming may lead you to feel. Sometimes people set schedules casually in some round number like six months or one year when no one would object to a more precise 7.6 months or 13.9 months.

Sometimes when a project reaches its end-of-schedule time, it turns out that no one seems concerned with the extra months the project then needs. Perhaps managers could have planned a little longer schedule in the first place.

Some executives feel profit pressures more keenly than time pressures. In circumstances such as these, you can take advantage of the time for effort tradeoff. After all, time may be important, but so is cost.

Table 14-2 lets you view the tradeoff numbers for systems of various sizes. Our intent is to show the improvements in effort (cost), peak number of people, and MTTD that are available for size of systems your organization builds.

Table 14-2. For each system size the effort, peak number of people, and MTTD are listed at the minimum development time and at 110, 120, and 130 percent of that time. We calculated this table at productivity index 12 and manpower buildup index 3. At other index values, of course, the figures will be different, but the pattern will be similar.

System Size and Characteristics	Development Time			
	Minimum	110%	120%	130%
10,000 SLOC				
Development Time (months)	7.04	7.74	8.44	9.15
Effort (person-months)	12.3	8.4	5.9	4.3
Peak People	2.7	2.5	2.3	2.1
MTTD (days)	3.9	6.2	9.6	14.4
25,000 SLOC				
Development Time (months)	10.42	11.46	12.50	13.55
Effort (person-months)	60.3	41.2	29.1	21.1
Peak People	7.9	7.2	6.6	6.1
MTTD (days)	1.8	2.9	4.5	6.7
50,000 SLOC				
Development Time (months)	14.02	15.43	16.83	18.23
Effort (person-months)	223.8	152.8	107.9	78.3
Peak People	23.0	20.9	19.1	17.7
MTTD (days)	0.9	1.5	2.3	3.4

75,000 SLOC				
Development Time (months)	16.69	18.36	20.02	21.69
Effort (person-months)	402.0	274.6	193.9	140.7
Peak People	36.7	33.4	30.6	28.2
MTTD (days)	0.7	1.2	1.8	2.7
100,000 SLOC				
Development Time (months)	18.88	20.76	22.65	24.54
Effort (person-months)	587.5	401.3	283.3	205.7
Peak People	47.9	43.6	39.9	36.9
MTTD (days)	0.7	1.0	1.6	2.4
250,000 SLOC				
Development Time (months)	27.95	30.75	33.55	36.34
Effort (person-months)	1,910.1	1304.6	921.2	668.8
Peak People	105.3	95.8	87.8	81.0
MTTD (days)	0.5	0.7	1.1	1.7
500,000 SLOC				
Development Time (months)	37.62	41.39	45.15	48.91
Effort (person-months)	4,656.9	3,180.7	2,245.8	1,630.5
Peak People	190.8	173.5	159.0	146.8
MTTD (days)	0.4	0.6	0.9	1.3
1,000,000 SLOC				
Development Time (months)	50.64	55.70	60.77	65.83
Effort (person-months)	11,353.6	7,754.7	5,475.3	3,975.2
Peak People	345.6	314.2	288.0	265.9
MTTD (days)	0.3	0.4	0.7	1.0

The truth about "reducing the cycle"

In recent years companies have tried mightily to reduce the length of the development cycle. They believed that getting a product on the market-place a few months sooner increased life-cycle profits substantially. In part, this drive for speed may have spilled over from the sense of urgency the Cold War brought. In part, widening commercial competition may have intensified it.

The push to reduce cycle time may carry a cost. In software, trying to beat the minimum development time is foolhardy. Even at the minimum development time a project will be much more costly than at a schedule 10 to 30 percent longer.

We are not arguing that getting to market sooner is not good in the abstract. We are pointing out that it involves tradeoffs. The methods of this chapter give you the means to quantify the cost of reducing software schedules to the minimum. They give you the means of assessing the savings that go with a little more time.

Reference

[1] T. DeMarco, *Controlling Software Projects*, Yourdon Inc., New York, N.Y., 1982. 284 pp.

Chapter 15

The Effects of Smaller Development Teams

"We have a three-year backlog of software projects," a vice president of operations of a telecommunications company complained. "We desperately need to get more of our operations automated if we are to stay in this fast-moving ball game."

"I'm trying as hard as I know how to catch up," the director of software development told us later. "Experienced developers are quitting faster than I can find people to start training. My friends in other companies tell me the same story. What can I do?"

"Use smaller development teams." That was our general answer to this common complaint. It sounds silly to most people at first. Of course if you have read the previous chapters, you realize there is a sound basis for this point of view. The experience of the software development division of a large telecommunications company illustrates the value of a smaller team. The division develops both business and communications software for other divisions.

A new telecommunications project generally involves land, building, electronic switches, land lines and microwave facilities, and staffing, as well as software. The software is usually a small part of the total cost. The cost benefits of getting the entire project online are substantial. Consequently, the company usually pushes projects of this type as hard as it can.

In this case the company has already issued the first release, 75,000 lines of C code. The process productivity index is 10, one point below average for a telecommunications system. The manpower buildup index is 6, a very rapid buildup, indicative of the company's desire to get new systems into operation fast.

This strategy takes lots of people and money, but the company was not unhappy with that. Senior executives felt it was important to keep up. Unfortunately, this strategy also resulted in a high incidence of defects when the software first became operational. The company was not at all happy about that aspect, so the managers were willing to listen to our "smaller teams" argument when a follow-on release came up.

The release was to be 15,000 lines of C code. Using the productivity index and manpower buildup index calibrated from the first release, we computed the planning numbers to be

Minimum development time:	7.7 months
Effort	128.0 person-months
Mean Time to Defect	0.74 day
Peak staff	26.0 people

Of course, there is only 50 percent probability of completing at these numbers. A check with the historic database showed that very few projects of this size and application type had occurred previously. It seemed wise just as a matter of statistical probability to consider a smaller team operating over a little more time.

We ran off a selection of the possibilities, listed in Table 15-1. Figure 15-1 diagrams the decline in effort. Figure 15-2 shows the improvement in MTTD.

Table 15-1. Using a smaller team, at the expense of taking a little more time, greatly reduces the effort and improves the MTTD.

Time Extension	Approx. Peak Staff	Time (Months)	Effort (PM)	% Effort change	MTTD (Days)	% MTTD change
1.00	25	7.7	128	0	0.53	0
1.05	19	8.1	104	-19	0.69	30
1.12	15	8.6	83	-35	0.92	74
1.21	10	9.3	60	-53	1.37	158
1.26	8	9.7	50	-61	1.73	226

Savings from Using Smaller Teams

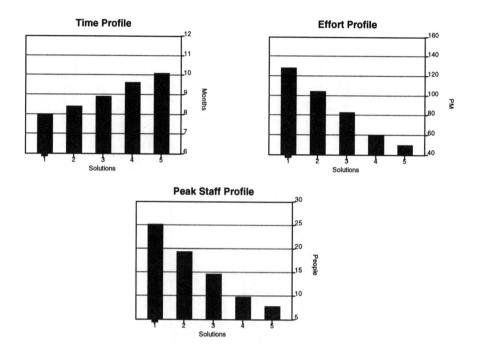

Figure 15-1. If this telecommunications organization moves to smaller development teams, it will save substantial amounts of effort and, of course, cost. It can make smaller teams feasible by extending the development time beyond the minimum.

Reliability Improvement from Using Smaller Teams

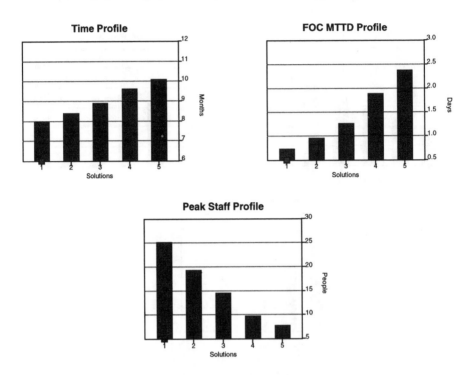

Figure 15-2. The same strategy results in a great improvement in MTTD. System operation during the initial period in the field will be more trouble-free.

We pointed out to the general manager of the software division that the increase in the development schedule for this range of smaller teams extended from a couple of weeks to a couple of months. Was a few weeks going to make that much difference to the users?

"If I ask them, they will say yes," he replied. "I know some organizations start projects with great fanfare and then sort of lose interest in most of them along the way. Sometimes a project just peters out altogether. Other times no one even notices when the project finishes several months late."

"Yes, an engineer we know once quit a Fortune 100 company because no product he had worked on in four years had reached the manufacturing stage," we recalled. "He said he had become an engineer because he wanted to build things."

"He should have come to work for us," the general manager said. "We try to think through what we need before we begin spending money on it. Then we follow through. Schedules are a serious business with us."

"With us, too," we agreed. "But effort is just as serious. For two months more, you can save about 80 person-months of effort. That is about $800,000 at your rates. With the people freed up this way, you could get started on two more 40-person-month jobs that you are holding for lack of people."

"You tempt me," he returned, "but tripling MTTD tempts me still more. The excess errors in the first month or two of operation provoke our users. They don't get the full benefit of the system for the first couple of months anyway."

Well, we had given the general manager the facts. He had to balance the gains against the intangibles; mainly, would the users be willing to hold off a couple of months for $800,000 and a lot fewer errors? (The using organizations paid the software division for their systems.) We heard that he talked to some of them.

At any rate, he did follow the new strategy. The numbers did work out as predicted. Now the division is making a continuing effort to persuade the using divisions to allow the software division to employ smaller teams. In return for a little more time the using divisions experience less trouble during early operations. The divisions are beginning to appreciate, too, that more than twice as many projects can be completed with the same number of software developers. It is a way to reduce their backlogs. That is better for their bottom lines than getting a lesser number of projects done a little faster. It's a win-win strategy.

In virtually any undertaking ... a very small fraction of the participants produces a very large fraction of the accomplishments. ...Increasing the number of participants merely reduces the average output.
—Norman R. Augustine [1]

PS. Geniuses are hard to get into the herd, but even without them, smaller is better.

Reference

[1] N.R. Augustine, *Augustine's Laws,* Penguin Books, New York, N.Y., 1987, 484 pp.

Chapter 16

Operating within Constraints

The more concisely and clearly boundary conditions are stated, the greater the likelihood that the decision will indeed be an effective one and will accomplish what it set out to do. —Peter F. Drucker, [1]

People responsible for business projects operate in a world of constraints, and they must plan a project in the light of those constraints. As we developed in previous chapters, the estimated size of the proposed system, the process productivity of the project organization, and the manpower buildup of the organization set limits within which planners must locate development time and effort.

In addition five constraints are often present: somebody's conjecture of the maximum budget actually available; a surmise of the maximum development time that external factors will permit; an estimate of the maximum number of suitable people that will be available for this project; an appraisal of the minimum number of people necessary to cover all bases sufficiently; and the desired mean time to defect.

Picking your battles

The unfortunate truth is you can't have everything. Everything in software development is connected to everything else. If you try to push one management number down, another goes up. At the time you are planning the schedule and effort for a project, you have a given size and process productivity (SLOC/PP). Manpower buildup is also fairly fixed. In both cases you can vary effort and development time, but their product must remain the same, because the SLOC/PP has not changed. Similarly, the ratio of effort and development time must remain the same, because manpower buildup remains the same.

There are limits on how far you can push effort and development time one way or the other. For example, the intersection of the Size/PP line and the MBP line sets the minimum development time. At this time, of course, the effort is maximum. In the other direction the limit is less definite, but it is seldom helpful to extend the development time beyond 130 percent of its minimum. At that point the effort is approaching a practical minimum. Because the relationship between effort and development time is exponential, the reduction in effort is less marked after that point.

So far we have been talking in terms of effort, that is, person-months or person-years. Effort (or its equivalent) can be expressed in several other forms, such as cost, maximum peak staff available, and minimum peak staff available. In considering an estimate, management might wish to see the effect of all these factors. That is what Figure 16-1 visualizes.

In this case, management or the customer has set four constraints:

❑ Maximum budget: the horizontal line across the diagram.
❑ Maximum peak staff: line rising from left to right, passing through the first dot.
❑ Minimum peak staff: line rising from left to right, passing through the second dot.
❑ Can't exceed date (maximum development time): vertical line at 12 months.

Solution Region Bounded by Constraints

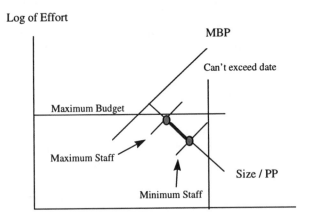

Figure 16-1. The range of acceptable solutions lies between the two dots. Under the constraints of this project, management may exercise judgment within this range.

What we have been doing is an application of a technique called linear programming. This technique draws lines representing the constraints on a log-log diagram. The logarithmic scales enable the lines to be straight (although the software equation and the manpower buildup equation are nonlinear). We know from the previous diagrams that the practical tradeoff region lies on the Size/PP line. The other constraints narrow a length of the line, between the two dots, within which management is free to trade time for effort.

The effort permitted by the manpower buildup line is within the cost constraint and the maximum peak staff constraint. The minimum peak staff line establishes a development time that is within the "can't exceed" date. Management can select a development time between the two dots while staying within the imposed constraints.

In this example there were solutions within the constraints. That is not always the case. Sometimes you must relax one or two constraints to reach a solution.

Optimizing the solution

What estimators really want, we have found, is a computing method that can accept a numerical value for each constraint, the desired probability of achieving that constraint, and the relative weight for each constraint. As output, the method would provide the actual probability of achieving each constraint. Table 16-1 illustrates an example.

Table 16-1. The estimator enters the constraint value, the desired risk probability, and the weight. The tool provides the actual probability (in boldface) that each constraint can be achieved. Cost and effort are essentially synonymous, so the estimator enters only one or the other, in this case, cost. Here the governing constraints are time and MTTD.

Constraint	Desired Probability	Weight	Actual Probability
Time: 14.00 months	75 percent	10 percent	**21 percent**
Effort: —	—	—	—
Cost: $2,800,000	75 percent	20 percent	**95 percent**
Minimum Staff: 5.00	75 percent	20 percent	**99 percent**
Maximum Staff: 15.00	75 percent	20 percent	**99 percent**
MTTD: 16.000 hours	90 percent	10 percent	**63 percent**

The constraints should be real, not vague hopes or fears. They should be related to the realities of the proposed project. At this point we already have an estimate of size. We have values for process productivity and manpower buildup. These set the limits within which the work must be carried out, as Figure 16-1 illustrates. It might be nice to have a product ready for the next annual show in your field seven months hence. But if the minimum development time is 10 months, setting a constraint of seven months is not realistic.

A 50 percent probability of meeting one of your constraints might not keep your stomach acids under restraint. You might desire a greater probability, up to 99 percent. When you indicate a development time of 14.00 months at a desired probability of 75 percent, you are saying that "I want to aim for a schedule for which I can be 75 percent certain that it will not exceed 14 months." That might mean planning a schedule of 13 months to be 75 percent sure that the actual schedule will not run beyond 14 months.

The six constraints may have different weights in various organizational settings, in different applications, or for different customers. In an embedded control application, in which safety is critical, MTTD might carry greater weight than schedule length.

Given all this input, the process of finding an optimized solution becomes one of first determining the range of schedules possible within the constraints and then, within that range, computing individual, weighted, and joint probabilities. Figure 16-2 gives a conceptual view of this process.

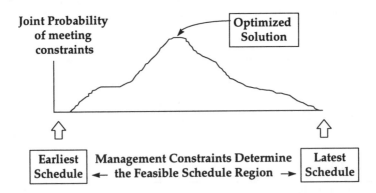

Optimized Solution
Conceptual View of Process

Figure 16-2. Somewhere between the earliest schedule and the latest schedule permitted by the constraints lies the greatest joint probability of meeting those constraints.

The results of this calculation are in the Actual Probability column of Table 16-1. The actual probability may or may not be as high as the desired probability. You may have set constraints and desired probabilities that cannot be achieved. "You can't have everything," as we said a while back. Unfortunately, reality continues to intrude. In that case you would have to modify your constraint values or your desired probabilities to get the best combination you can. When these voluminous computations are computerized in a tool, you can examine many possibilities in a reasonable time span.

Dealing with Windows-NT constraints

Will Rogers used to say, "All I know is what I read in the newspapers." Well, we know a few things the media hasn't picked up on yet, but in this case all we know is what we read in the *Wall Street Journal,* May 26, 1993. G. Pascal Zachary provided a few numbers about Microsoft Corp.'s development of its Windows NT program, which was then about eight months behind its current schedule. It is a huge program, "a staggering 4.3 million lines of code," wrote Zachary. It cost "more than $150 million to develop." It employed "200 developers and testers." On the basis of these "facts," we estimated a process productivity (PI) index of 18.0 and a manpower buildup index of 2.2.[*]

Given "constraints" of this nature, we thought it would be instructive to see what Microsoft was up against six or seven years ago when it started this project, or its predecessor. Table 16-2 lists the constraints we selected.

The QSM software tool provided the solution listed in Table 16-3. The table shows that there was a good chance of completing the project in about six or seven years—the time now elapsed—but the original time projection was less than six years. The project has now slipped its schedule several times.

The table shows that there was a poor chance of completing the project at the constraint level of effort. In fact, the tool had to increase the level of effort a little, but the odds were still poor.

---

[*] These "facts" differ from those guesstimated by other observers. Therefore, we reiterate that our Microsoft example is purely hypothetical.

Table 16-2. We consider these constraints applicable to a project of the scope of Microsoft's Windows NT.

Measure	Constraint	Desired Probability	Weight
Time (months)	80.00	90 percent	16
Effort (person-months)	10,000.00	90 percent	16
Cost	$200M	90 percent	16
Minimum Staff	100.00	90 percent	16
Maximum Staff	250.00	90 percent	16
MTTD (hours) at First Operational Capability	8.00	90 percent	16

Table 16-3. This project cannot satisfy all its constraints at the desired probabilities.

Measure	Expected (50% value)	Actual Probability
Time (months)	76.26	83 percent
Effort (person-months)	10,439.53	41 percent
Cost (millions of dollars)	138.823	99 percent
Minimum Staff	200.00	100 percent
Maximum Staff	200.00	92 percent
MTTD (hours) at First Operational Capability	6.91	19 percent

The reliability situation is still worse, but we defer talking about it to Part IV.

"I remember skittering by Monte Carlo simulation and linear programming in college," Phil told us, *"but I never expected to actually use them."*

Now you can, we replied. You don't even have to formulate the problem for a generic Monte Carlo or linear-programming program. Software estimating packages put it in a form that fits the software process.

Reference

[1] P.F. Drucker, *The Effective Executive,* Harper & Row, New York, N.Y., 1966, 178 pp.

Chapter 17

Bounding Risk

I'm shocked, shocked to find that gambling is going on in here.
—Claude Rains in *Casablanca*

What if they gave a war and the software did not come? Large military procurement programs heavily dependent on software exceeded their schedules by 20 months on average, according to a study of 82 projects by US Air Force Colonel Joseph Greene, Jr. (By the way, 20 months is about half the length of US participation in World War II.) Projects less dependent on software overran by only seven months.

"The department is paying a huge penalty for not dealing with its software problems," said Greene, now retired. "The penalty is not just late software—it is degraded war-fighting capability" [1].

If there had been such a thing as software in the 1940s, the war might have ended before the software was ready.

Or in entrepreneurial circles the market might have gone to some one else before your software was ready. It would help to steady your nerves when you are about to finalize the effort and schedule estimate if you could check it against industry-wide experience. It would also help if you could visualize your risk situation.

Consistency check

As we described in earlier chapters, Quantitative Software Management, Inc. has a database of 3,885 systems divided into nine application areas. The database lets you compare your current estimate against the average of the historical projects in the same application area. Table 17-1 illustrates. The table is based on the Windows NT project described in the previous chapter. The last column indicates the number of standard deviations by which the current estimate exceeds or falls below average.

Table 17-1. Checking the management numbers that represent your estimate against corresponding numbers in the database gives you some assurance that your estimate is within the bounds of reality.

Management Number	Database Average	Current Estimate	Standard Deviation from QSM Average
Productivity Index	11.9	18.0	1.50
Size (SLOC)	N/A	4,000,000	N/A
Time (months)	75.11	87.07	0.25
Effort (person-months)	14,553.06	6,146.43	-0.80
Uninflated Cost (millions of dollars)	N/A	56.347	N/A
Average Staff	193.75	70.60	-1.03
MTTD at Full Operational Capability (days)	34.97	1.67	-2.05

The Microsoft project is doing very much better than the average systems software project in the database.

Key variables from the table are translated into visual form in Figure 17-1. Diagonal lines on each log diagram show the spread of the historical data. The center line is the average, bordered by plus-and-minus one-standard-deviation lines. The black squares show the location of the current estimate. Development time, for example, is about a quarter standard deviation greater than average, a fact Table 17-1 also shows.

Risk profiles

The QSM software tool carries out nominal computations in terms of expected values—50 percent probability of success. The risk-schedule profile of Figure 17-2 lets you evaluate the degree of risk associated with your current estimate of planned development time. Planning to carry out the project in 76 months gives you a 50 percent chance of completing in that period. Working to this same plan, you have an assurance of 75 percent of completing within 79 months. Bidding only a few months longer than your planning schedule greatly increases the odds of successful completion within your bid time.

Achieving nearly complete certainty of success, however, requires you to allow many more months. The profile curve is nearly linear out to about 85 percent assurance. Beyond that point, it veers sharply upward and is rising nearly vertically in the upper 90s. In other words, it is difficult to pick one number of months that assures near certainty of success.

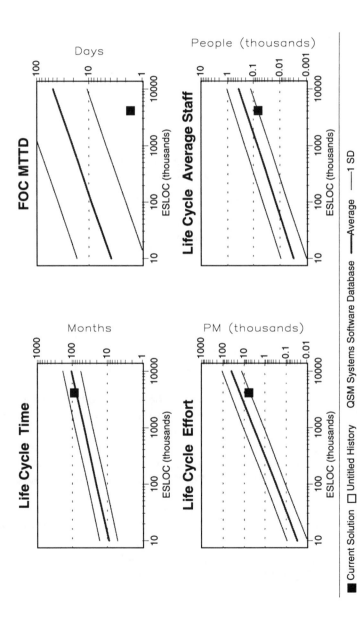

Figure 17-1. Life cycle development time, effort, average staff, and MTTD at full operational capability are shown by black squares on a field of average and standard deviation lines representing the historical database.

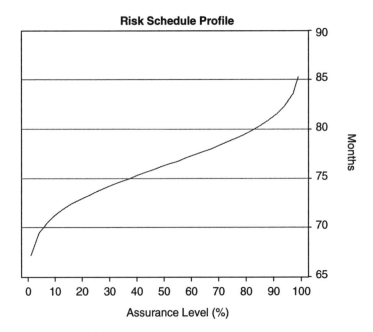

Figure 17-2. On the basis of a given planned development time, the risk schedule profile plots the probability of successful completion against months of actual development time. For example, on a planned schedule of 76 months, there is a 90 percent probability of completing within 81 months.

The same phenomenon appears at the low end of the curve. If you plan to carry out the project in 76 months, you have a decent probability, like 20 to 40 percent, of completing a few months early. But the chance of completing many months early is very small. Again the curve veers to the nearly vertical.

The software tool can also break down the risk-schedule profile by project milestones, as Figure 17-3 shows, and it can print out tables listing the numbers that correspond to the profiles.

The tool can provide similar risk profiles and tables in terms of event dates instead of number of months. Also, in addition to providing these profiles and tables for schedule, it can produce them for effort, cost, and MTTD.

Although you can minimize and statistically describe risk, you cannot eliminate it. The future is unknowable. The sky may fall in, as Chicken Little proclaimed, before the project is completed. But we are not talking about the sky or a thousand other possibilities. We are talking about effort and schedule.

Risk Schedule Profile by Milestone

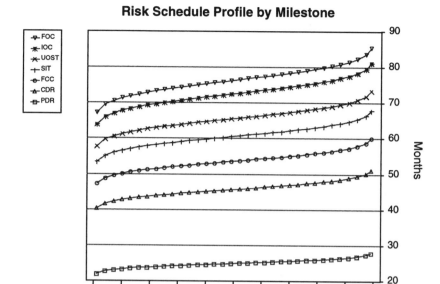

Figure 17-3. The expected value of the planned development time at the first milestone is 25 months. The number of months rises until full operational capability at 76 months. The milestones are

IOC	initial operational capability
FOC	full operational capability
SIT	system integration test
FCC	first code complete
UOST	user-oriented system test
CDR	critical design review
PDR	preliminary design review

In that limited arena software people *can* minimize risk. A schedule longer than the minimum development time is less risky than one at or below the minimum time. A plan set within realistic constraints is less risky than one outside one or two constraints. A plan that compares favorably with the historical experience of similar projects is less risky than one outside that ball park.

Even when managers minimize the schedule-effort risk using these practices, plenty of risk remains. The various risk profiles—schedule, effort, cost, MTTD—let you assign probabilities to the remaining risk. The intent is to give you the tools to select the degree of risk that best fits your situation. And much of your situation is imponderable. The risk profiles are an aid to judgment.

Reference

[1]   E. Williams, "The Software Snarl," *Washington Post*, Dec. 9–12, 1990.

Chapter 18

Planning the Work

Businesses operate on commitments, and commitments require plans.
—Watts S. Humphrey [1]

The next task is to turn the estimates of total effort and schedule into a distribution of effort and cost over development time. This distribution follows the pattern of a Rayleigh curve, as we noted in Chapter 3.[1] Several versions of the Rayleigh curve can be generated: staffing, cost, and code production. (We defer defect and MTTD curves to Part IV.) The software tool that performs these chores can produce 174 reports and presentations, either on screen or on paper. We limit ourselves here to a mere indication of the variety you can employ.

QSM's Software Life-Cycle Model tool (SLIM) computes these curves and tables on the basis of the current estimate you have selected. If the plans that it presents don't match your needs, you can go back to estimating mode and select a different estimate. The tool automatically recalculates the presentation material to match the new estimate of effort, cost, and development time.

---

[1] The Rayleigh rate equation, expressed in terms of effort and development time, is:

$$y' = (K/t_d^2) \, t \, \exp(-t^2/2t_d^2)$$

Once estimators have established values for effort and development time, K and t_d, the values of y' over time t can be computed. If values one standard deviation above and below the expected values of effort and development time are substituted in the equation, boundary lines, as shown in Figure 3-5, can be added.

In cumulative form the equation is:

$$y = K \, [1 - \exp(-t^2/2t_d^2)]$$

Reports and presentations

Reports and presentations are ever present in a manager's life. Fortunately, most of the data can be formatted and generated automatically.

Staffing plan. Figure 18-1 shows a projection of the people required to implement the Windows NT estimate. The figure is based on expected values at 50 percent. It shows two phases: requirements and high-level design, and construction and test (or main build). It also shows about a year of overlap between the two phases, but the phases might also be separated by a specified gap.

Variations of Figure 18-1 include (1) values of effort and time other than the expected 50 percent values; (2) an aggregate curve, in which one curve totals the two curves in this figure; (3) separate diagrams for each phase. You can also separate the requirements and high-level design phase into two separate curves: requirements and high-level design. In Figure 18-2, we have broken the construction and test phase into six classes of work. The numbers corresponding to points on the curves are available in tables.

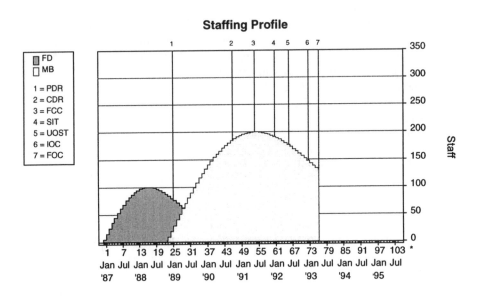

Figure 18-1. Microsoft might have staffed the Windows NT project along these lines. FD stands for functional design; MB for main build. The milestones are as follows: PDR is preliminary design review, CDR is critical design review, FCC is first code complete, SIT is system integration test, UOST is user-oriented system test, IOC is initial operational capability, and FOC is full operational capability.

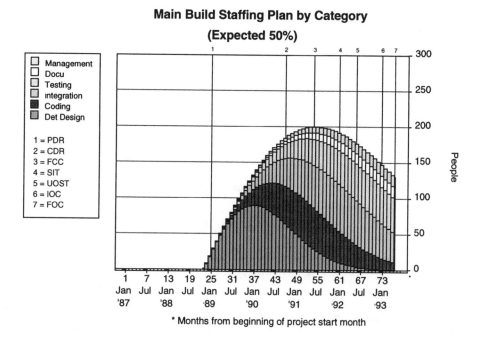

Main Build Staffing Plan by Category
(Expected 50%)

* Months from beginning of project start month

Figure 18-2. The construction and test phase can have separate curves for,detailed design, coding, integration, testing, documentation, and management. This main build staffing profile was drawn at expected, or 50 percent, values. Milestone abbreviations are explained in the caption for Figure 18-1.

Figures 18-1 and 18-2 are rate curves; that is, they reflect the number of people per time period, such as months. You can also express this information in cumulative form, as Figures 18-3 and 18-4 illustrate.

Cost plan. Because cost is merely effort multiplied by a factor, you can generate a similar set of diagrams in terms of cost. The cost may be uninflated or may be increased year by year to reflect an estimate of coming inflation.

Bar and Gantt charts. You can also present the information as bar charts, drawing each breakdown at various assurance levels, such as effort breakdown, cost breakdown by category, effort between milestones, and cost between milestones.

Figure 18-5 is a Gantt chart showing the time assigned to each project phase at various assurance levels. Many people are accustomed to this form of presentation. However, Gantt charts do not effectively portray the rework inherent in software development.

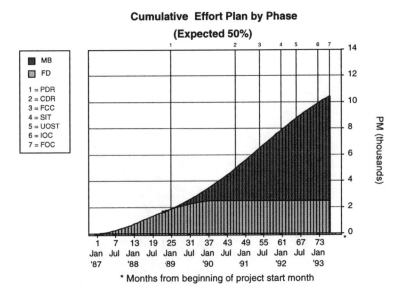

Figure 18-3. In cumulative form you can weigh the effort expended to date against the total effort estimated. FD is functional design; MB is main build. Milestone abbreviations are explained in the caption for Figure 18-1.

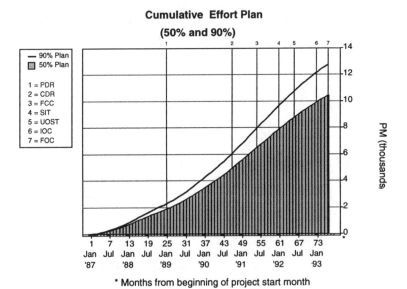

Figure 18-4. Here both the 50 percent expected value and the 90 percent value are drawn on the same diagram. The extra effort shows the margin required to be 90 percent certain of completing the project in the time estimated. Milestone abbreviations are explained in the caption for Figure 18-1.

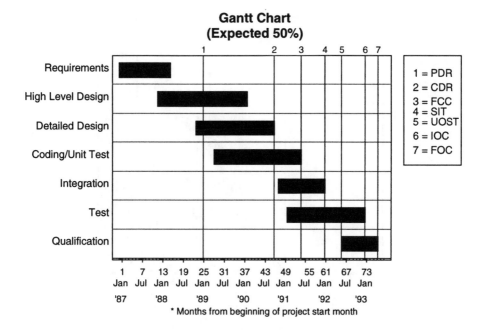

Figure 18-5. Gantt charts show the expected start date and end date for each phase: requirements, high-level design, and so on. Milestone abbreviations are explained in the caption for Figure 18-1.

Code production. The software tool projects code production at four levels, as Figure 18-6 illustrates. Code production is rapid during the early part of the schedule but tails off during the last year, supporting the common allegation that software is 90 percent finished for a long, long time. The valid-product curve does not reach 100 percent at full operational capability or delivery time. In a project of this magnitude a substantial number of defects would ordinarily remain to be found and corrected after delivery.

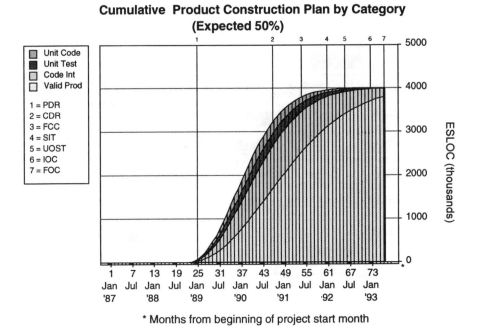

Cumulative Product Construction Plan by Category (Expected 50%)

Legend:
- Unit Code
- Unit Test
- Code Int
- Valid Prod

1 = PDR
2 = CDR
3 = FCC
4 = SIT
5 = UOST
6 = IOC
7 = FOC

ESLOC (thousands)

1	7	13	19	25	31	37	43	49	55	61	67	73
Jan	Jul	Jan	Jul	Jan	Jul	Jan	Jul	Jan	Jul	Jan	Jul	Jan
'87		'88		'89		'90		'91		·92		'93

* Months from beginning of project start month

Figure 18-6. The cumulative curve shows the quantity of code that has passed production and test points.

Unit code:	reported as coded
Unit test:	passed unit test
Code int.:	passed integration
Valid prod.:	code that has passed system or acceptance test.

Milestone abbreviations are explained in the caption for Figure 18-1.

Documentation. Software organizations devote a large portion of project time and effort to documentation. They like to have an estimate of the number of pages to be produced in each phase of the project, such as requirements, design, development plan, test plan and reports, and user manuals. From the page estimates they can judge the amount of effort and time needed.

Because there is great variation in documentation requirements from organization to organization, a single documentation algorithm, to be used by all organizations, is not practical. Organizations must be able to tune the algorithm to fit their own situation.

Most important, they must be able to tune the algorithm to reflect how intensive the documentation requirements are in their own case— little formal documentation; user manuals only; or full documentation of all development phases. The degree to which you use documentation tools affects the algorithm—no specialized tools, no formal templates; loosely integrated tools necessitating some rekeying, some templates; high use of integrated tools, minimizing rekeying and reformatting.

A broad plan

Although the SLIM tool breaks down staffing, cost, code production, and so on by categories or between milestones, its presentations are essentially at an overall project level. You must continue to prepare work breakdown structures, to assign subsystems to individual designers, modules to programmers, test plans to testers, and so on.

You will find these project-level reports and presentations useful for at least three purposes: (1) foreseeing the need for staff and funds; (2) planning the transfer of staff from projects that are winding down to those that are starting up; and (3) tracking and controlling progress on the project.

With an appropriate set of curves and tables you can see months ahead the project requirements for people, space, equipment, and funds. You can take timely action to meet these needs.

On an organization-wide basis, by aggregating these curves and tables for all projects, you can see whether staff coming off projects that are winding down will match the needs of projects that are building up. In fact, you can even forecast the needs of the software development staff far enough ahead to take timely action, including to plan additional hiring or training; to seek further work for those who will become available; or even, in dire times, to make negative plans.

Using these staffing, cost, and code projections to track and control the progress of ongoing projects is the subject of the next few chapters.

Reference

[1] W.S. Humphrey, *A Discipline for Software Engineering,* Addison-Wesley, Reading, Mass., 1995, 789 pp.

Chapter 19

Working the Plan

"Coding ... is '90 percent finished' for half of the total coding time."
—Fred Brooks [1]

Richard A. Zahniser expressed much the same thought as the 90-90 rule: "The first 90 percent of any systems work takes 90 percent of the time. The last 10 percent [also] takes 90 percent of the time. The 90-90 rule sounds like a joke. It's not" [2].

Dynamic control can cure the 90-percent-complete syndrome. The idea is to have real indicators of how much work you have done, rather than insubstantial hopes. Then you pay attention to the real measures.

You very likely already know how to dynamically control a production process. You may already be using dynamic control in some of your operations under the name "statistical process control."

Here are the basic principles:

❑ Pick a few key factors that reflect the progress of the work.
❑ Figure out how to measure these factors.
❑ Project the numerical value of these measures over the proposed project schedule.
❑ Compare the actual values with the projected values.

> If the actuals are close to plan, smile.
>
> If the actuals deviate from plan, frown.
>
> If the actuals are way off, act.

"Your assignment, ladies and gentlemen. Find 100 percent and bring it
to me."

Pick key factors

In previous chapters we picked as key factors the management numbers:
development time (schedule), effort (person-months or cost), functionality (size in source lines of code or function points), and reliability (defects
per time period or its reciprocal, mean time to defect).

Development time, of course, is the time scale along which the other
factors operate.

Effort is an input, not an output reflecting the progress of the work.
It is the input that does the work that produces the output. Nevertheless, it is a factor to be controlled. If a project is not putting in the
planned effort, no one should be surprised if it fails to produce the
planned output.

The other two factors, functionality and reliability, are indicators of
the output. In plainer words, code production rate is a measure of the
quantity of work. Some organizations have preferred to measure functionality in terms of function points, modules, or subsystems completed,
rather than source lines of code. We find, however, that the function-
point method, being a logical expression of function, does not lend itself

well to tracking. The physical expressions—lines of code, modules—are better for that purpose.

The defect rate is an indicator of the work's quality level. As rates per time period, the numbers provide continuous feedback of production rate and quality.

Production rate and quality—these two factors are central to every production process. At the risk of mouthing platitudes, the first has a lot to do with the price the ultimate user has to pay. The second bears on the satisfaction that the user experiences in using the product. We deal further with quality in Part IV.

In addition to these key factors, many organizations have tracked the dates on which project phases (requirements, high-level design, and so on) and milestones (preliminary design review, critical design review, and so on) were passed. Others have counted the number of pages of documentation completed, number of reports completed, and the like.

Measure key factors

How to measure most of the key factors is clear, even though there may be difficulties in doing so. Finding a way to represent the amount of work done, however, has been perplexing. All the brainpower devoted to gathering information from prospective users, analyzing requirements, writing specifications, and creating design finally finds measurable expression as source lines of code.

Some view SLOC primarily as a measure of coding effort, not as a measure of the rest of the software process. Still, it is the last human-generated step in the software process before a machine process compiles the code. In that sense it represents the output of human effort in the software process.

That may be the reason SLOC is the commonly used measure of software work. Function points have seen some use, particularly in business software. A few organizations have counted larger chunks of software, such as modules and subsystems.

But SLOC is the most practical measure we have. As a tracker of progress, it works. As Peter Drucker says, "In the last analysis management is practice. Its essence is not knowing but doing. Its test is not logic but results. Its only authority is performance" [3].

Project key factors

Planners project key factors against time. The projections may be in rate or cumulative form. Figure 19-1 is an example of a staffing projection in

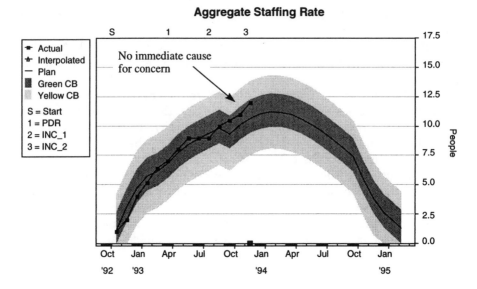

Figure 19-1. The staffing-rate control diagram plots people over development time in relation to plan (solid line) and green and yellow control bounds. Staffing rate means the number of people working on the project during a time period.

rate form. The solid line, Plan, extending across the diagram, represents the likely level of accomplishment. Because the historical data behind the projections and the projections themselves are uncertain to some degree, we also draw in two sets of upper and lower control bounds. The green control bounds (on a color screen) are a close-in set. The yellow control bounds are farther out. The black squares represent actual staffing during the first 14 months of the project. Since actual staffing is close to plan and within the green control bounds, managers have no cause for concern so far in the project.

Previous diagrams have carried numbers at the top that correspond to US Department of Defense milestones, such as CDR, Critical Design Review. However, many software organizations have their own milestones, and this diagram incorporates one such set: INC_1 and INC-2. In this case the acronyms stand for increments in an incremental build development. In general, these milestone markers are tailorable in definition, acronym, and placement (within limits).

Figure 19-2 projects code production in cumulative form. You can present other measures of functionality in similar diagrams.

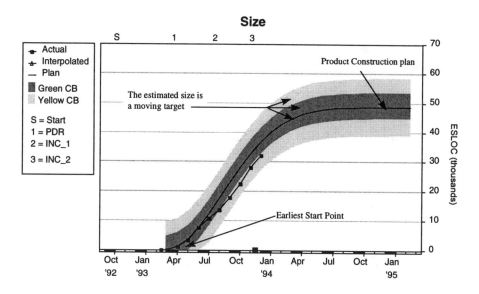

Figure 19-2. We expect the code production rate to fall near the center line or, in most cases, within the control bounds. In this example, code production, in the yellow area, is less than the production expected by the plan.

Measure actuals

In Figure 19-1 the actual staffing numbers for the first 14 months of the project are blocked in as black squares. The actuals are running close to the projection, usually within the control bounds. Four months later, as Figure 19-3 shows, actual staff is substantially exceeding the planned number. The managers concerned should look into the reasons.

Compare actuals against the plan

Perhaps some one is parking surplus people here; that may be necessary, but the project will likely exceed its effort estimate later. Perhaps work has been broken out for additional people earlier than planned; as a result, the project is going well and will finish early within the effort estimate. Perhaps the work is turning out to be more difficult than expected; the people are needed all right, but effort and schedule are likely to be overrun. It is time to find out what is going on and to act, if necessary.

So long as the actuals fall within bounds, the project is proceeding more or less according to plan. People say, "The project is in control." You don't have to worry about it.

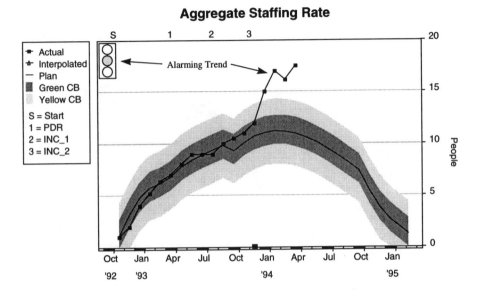

Figure 19-3. After the 14th month the staffing rate begins to exceed the upper control bound.

We think it is a good idea for executives, managers, and supervisors to look at the control charts every month, just for their personal edification, even though they don't have to do anything about projects that are in control. It makes everyone happy to think that management is interested. For managers it's a lot like a "Maalox moment," and it takes only about that much time.

Uh oh! Sometimes the actuals fall outside the bounds. If you are a general executive, not a software technologist, you aren't supposed to know all the nitty gritty about what could be done. Of course, somebody should know and they should be doing something. Then, if what they tell you seems convincing, you can have another "Maalox moment."

It's in the doing something that dynamic control gets sticky. You might not have learned that in kindergarten, but you've picked it up along the way.

Fred Brooks points out that "every boss needs two kinds of information, exceptions to plan that require action and a status picture for education." Then he "must discipline himself not to act on problems his managers can solve, and never to act on problems when he is explicitly reviewing status" [1].

Most of the time software people can make a few technical corrections and get the project back on plan. Once in a while a deviation is so serious that it reaches higher levels:

❑ People may say the project has to be replanned. That extends the schedule and increases the cost. It affects the customer. And that affects higher levels.

❑ People may say that to meet the schedule the product's functionality has to be reduced. Again, that involves the customer.

❑ People may throw up their hands. The project is totally out of control. See next chapter. They want to cancel the project. It's time for Maalox.

A note on the mechanics

It would be a considerable amount of manual work to draw control diagrams every month like those in this chapter. In fact the amount of work is so great that it often discourages organizations from even attempting to control the rate of progress, or at the very least greatly limits the amount of control they do attempt.

If you need to do a lot of planning and control, it's best to delegate the job to a program. SLIM-Control, part of the QSM tool suite, can graph and provide backup tables of

❑ Cumulative SLOC or other expressions of functionality;

❑ Staffing rate, or people per month;

❑ Cumulative effort;

❑ Cumulative cost;

❑ Phase progress;

❑ Milestone progress;

❑ Gantt chart;

❑ Percentage complete;

❑ Up to five user-defined metrics, such as pages of documentation.

References

[1] F.P. Brooks Jr., *The Mythical Man-Month: Essays on Software Engineering,* Addison-Wesley Publishing Co., Reading, Mass., 1974, 195 pp.

[2] R.A. Zahniser, "A Massively Parallel Software Development Approach," *American Programmer,* Jan. 1992, pp. 34–41.

[3] P.F. Drucker, *Management: Tasks, Responsibilities, Practices,* Harper & Row, New York, N.Y., 1973, 839 pp.

"Farrell, just block in the model's general shape—We're behind schedule."

Chapter 20

"I Always Felt Frustrated"

This chapter is based on a consulting assignment performed by Law-rence Putnam. The names of the government agency and the contrac-tors have been changed.

A contracting agency can use a project forecasting system to learn the likely outcome of a contractor's bid. One outcome I did not anticipate as I prepared to perform a consulting assignment for the Federal Systems Agency was frustration. Yet frustration was my eventual feeling when I could not get enough good data to convince the FSA that its transporta-tion control project was heading for trouble.

The FSA had already spent years planning the project when I be-came involved in 1985. The agency had solicited proposals and wanted me to prepare a "should cost" estimate to guide its evaluation of the bids.

The agency had specified a three-year schedule. The two bidders were the Transport Systems Company and Radaran Corp. The pertinent facts in the bidders' proposals were

- ❑ Size of control software. 60,000 to 70,000 SLOC
- ❑ Language to be employed. Transport Systems Company proposed to use Pascal. Radaran Corp. proposed to use C.
- ❑ Microprocessor operating system. 5,000 additional SLOC (A real-time operating system for the then-new Motorola 68020 micro-processor already existed. Transport Systems Co. estimated 5,000 new lines would be required to adapt it to this application.)

The FSA had no information on the bidders' process productivity. My "should cost" estimates would be much more accurate if we could get his-torical data from the bidders on several of their recent projects.

In reply to my request, the FSA contract administrator said, "Oh no, we can't do that. Regulations don't allow us to ask vendors for information that would cost them money to compile."

In my view the agency lacked information that would make it an intelligent builder of a difficult system. The cost of this information would have been trivial compared with the costs that ensued. So I had to make my initial estimates on the basis of the industry-average productivity index, about PI 6 in this case.

First estimate

Each bidder proposed to complete the system, including high-level functional design, within the three-year time frame specified by the agency. One planned to build up people rapidly to a maximum of 65, the other even more rapidly to 100. Both planned to complete functional design in the first eight months.

Even with this limited information, I saw that the three-year schedule was unrealistic. The disparity between the proposed fast buildups and a more efficient slow buildup especially bothered me. I attempted to guide the agency to a more realistic schedule. With limited data, however, I could not be very positive in my objections.

I also tried to get the FSA to give me some feel for how long it expected the operational software to run between failures. It had specified a hardware MTTF of thousands of hours, a figure totally inconsistent with what people were achieving in real-time software.

The award went to Transport Systems Co., the one that had proposed to code in Pascal. In fact, its proposal had included the resumes of many experienced Pascal programmers. After the award, however, it decided to work in C, the language Radaran had proposed. At that point C was a new language. The people Transport had to bring in were mostly fresh from college where they had studied Fortran or Pascal. They had no experience in C.

Functional design

As they had proposed, Transport did throw people at the project. Staffing was up to 45 people within two or three months and up to 50 or 60 by six months.

At eight months, the point at which the schedule called for functional design to be complete, the company took a first cut at a preliminary design review. The FSA rejected a substantial part of the company's functional design as inadequate.

Functional design continued for another four or five months, iterating with feedback from the agency. The best guess I can make for the actual accomplishment of the preliminary design review is something like 12 to 13 months.

By this time Transport's opinion of the amount of work it would have to do on the microprocessor operating system was 25,000 to 30,000 SLOC, up from the original 5,000 SLOC. The size estimate for the other major functional pieces had nearly doubled from 65,000 to 125,000 SLOC.

Coding

Before the project had gone very far with detailed design, it began to write code. According to the plan, it was to write and integrate code in a five or six month time frame. Well, it did keep up the aggressive pace for about two months, doing the easy stuff straightaway. Soon coding ran ahead of the detailed design. The coding rate slowed drastically.

By this time I had done several tracking and control updates. Obviously, the schedule had already slipped. High-level functional design had been four or five months late. Now coding was slowing down and no integration was taking place, despite assertions by the contractor.

My analysis pointed to a schedule slippage of nine to 12 months. The Transport people did not accept this forecast right off. It usually took them three or four months to acknowledge their difficulties and propose a revised plan.

The coding rate kept slowing down. The rate would spurt up, then lag when it turned out that the code was unsatisfactory and had to be thrown away.

Two years into the program, the size estimate had grown to 180,000 or 185,000 SLOC. Code was less than 50 percent written. Transport had integrated nothing. Because of the rapid buildup, the project was running out of money. The FSA was keeping it going from its contingency funds.

In the meantime a conglomerate, Multiple Systems Corp. had acquired Transport. Multiple's system managers were running in and out of the situation. Ultimately the FSA exhausted its funds and Multiple began spending its own money. The contract committed it to deliver a system.

In periodic updates I advised the agency of where the project stood. Each time the schedule estimate was nine to 12 months longer than the contractor's current estimate, with a correspondingly higher cost and effort estimate. The agency did not want to hear the bad news.

The people I was dealing with seemed willing to listen to projections two or three months beyond whatever the current number was. They didn't like to hear nine to 12 months. The reason seemed to be that they could placate senior management and Congress on incremental slippages, but big slippages set off big alarms.

Nearing the end

I haven't touched the project for about two years—it is now six years old, double the original schedule. The last time I projected completion at the six-year point, but I hear it has six or eight months to go. The latest size estimate is 300,000 SLOC, about five times the initial estimate. Multiple Systems Corp. has been spending its own money these last two years. Altogether, it is not a pretty picture.

I had the feeling that the agency was having great difficulty managing a program of this size. They knew they were in trouble. They had lots of meetings. They never had a good feel for where they were or what managerial steps were necessary to get them back in control.

The vendor, too, knew it was in trouble, but it never knew just where the problems were or how they would affect the schedule. Its plan was faulty; It hired junior people; its managers were not as skilled as they should have been. It threw on too many people up front when there was no need for them at that point. Its obvious intent was to get the cash flow on the books.

Moreover, I fear that the delivered system is not going to do what it should. Defects will no doubt be numerous. If it doesn't work as well as it should, people's lives will be at stake.

Finally, I am not happy with my own role in the debacle. My feeling all along was that prospects of success were poor. I never really had enough data at any point to make a good estimate. *I always felt frustrated.*

Chapter 21

Shining Shadow Loses its Luster

This case is recounted by Douglas T. Putnam, also a QSM consultant. The details are disguised.

It was the summer of 1989. I telephoned my friend, Captain Robert Kramer of the Military Procurement Division. I asked him if I could demonstrate SLIM-Control on a software project that was under way.

"We are overseeing several hundred projects," Kramer responded. "I'll look through the Contractor's Cost-Schedule reports for a typical one."

A few days later he called back and said he had a candidate, Shining Shadow. It was a system in the Command, Control, Communications, and Intelligence category involving both customized hardware and software. The developer was a small business under a fixed-price contract. The software portion, consisting of an estimated 225,000 SLOC, carried a $17 million tag. The beginning date was October 1986, and completion was scheduled for January 1991.

Summer 1989

A couple of weeks after the call a package of raw data arrived and I settled down to examine it. The amount of information was encouraging: dates of project milestones, expected staffing for each month, and expected code production each month.

In this business you are lucky if people keep track of their data on the back of an envelope. Here we had regular monthly reports.

The reports listed actual staff assigned and code produced for each month through August 1989. The project had just passed the detailed design milestone, two months behind schedule. Apparently they were trying to catch up. They were up to 80 people, against a planned number of 50 to 60 during 1989. In fact, they had been over the planned staffing level for five months. *[Figure 21-1 shows this situation.]*

This staffing level required a monthly spending rate of $670,000. That is not a trivial rate placed against a budget of $17 million facing 18 more months of expenditure. I added up the person-hours expended to date, and at its burdened labor rate, about $100,000 per person-year, found the project had already spent about $14 million.

At this point coding had been under way for only a few months. Some 37,500 lines had been taken under configuration management.

I entered this data in my portable computer. With about two and a half years of actual data, the program computed the actual process productivity and manpower buildup index. With that knowledge, I could project development time, effort, and cost to the end of the project.

The result was alarming. Completion would slip 15 months, to May 1992. The cost overrun would be $17 million, or 100 percent.

Well, that would not be very good news for Captain Kramer. I decided not to make an issue of the results. I would focus on the value of a tool that can disclose problems early when management still has time to act.

Staffing Performance

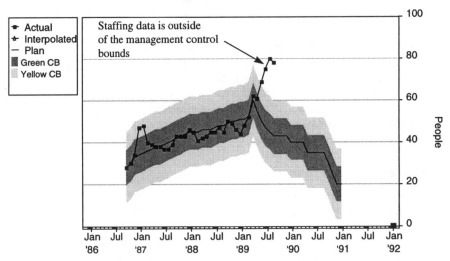

Figure 21-1. The solid line shows the staffing rate, taken from Shining Shadow's original plan. I added the upper and lower control bounds. Actual staffing numbers for recent months (black squares) indicate that the project is well out of planning bounds.

Briefing the program manager

In the Captain's office a few days later, he merely glanced at the thick pile of papers still in my briefcase. He said, "We're meeting the program manager, Colonel Roper, and his people in 10 minutes. You'll have 30 minutes."

That set my ticker pumping. I had never met these people. They had no knowledge of our methods. I had no briefing slides with me. And the gloomy prospect for Shining Shadow would put them in an unreceptive frame of mind, to say the least.

Captain Kramer located a manual on one of our products that had some figures I could put on a viewgraph for a quick briefing. I set up my computer and a device that projected the computer screen onto the wall.

A dozen people came into the briefing room. You could feel the tension. Last came Colonel Roper. The rest jumped to their feet. With an autocratic gesture he motioned them to sit. My ticker spurted again.

After running through seven background viewgraphs in seven minutes, I moved into the Shining Shadow material. I observed that they had already spent $14 million of the $17 million budget. I guess that wasn't the most tactful place to start. I had assumed that they were aware of such an obvious fact. An analyst for the nonprofit support agency on the contract jumped up almost immediately and claimed that the project had spent only $7 million.

I took him through the drill of adding up the person-months so far and multiplying by the burdened labor rate (which we agreed upon). He sat down, but he had taken up valuable minutes. His group must have been months behind in posting their figures. The analysts clearly felt threatened by what I was doing.

I plowed on. I tried to explain how I arrived at the 15 month schedule slip and 100 percent cost overrun. Now the project manager, a young captain, went crazy.

"What productivity assumptions are you using?" he demanded.

"I'm not assuming anything," I replied. "I derive a process productivity index from your own data to date. I use that value to project effort and cost into the future." I had the computer put up a chart. [The chart is similar to that in Figure 21-2.]

"Your process productivity, as taken off your own data, is less than you thought it would be when you planned this project three years ago," I continued. "Probably the work is more difficult than you expected. I based this projection on staying at your current peak staffing of 80 people. It assumes no reduction in the planned functionality or size. Unfortunately, to complete the work requires roughly 1,000 more person-months of effort than the original plan."

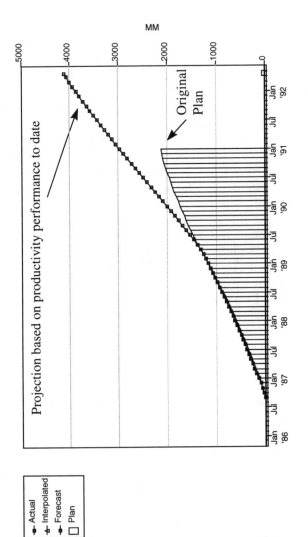

Figure 21-2. The bars represent the original plan. The black squares are actual cumulative effort to the summer of 1989. Thereafter, white squares represent cumulative projections to the spring of 1992.

I had the computer put up another chart. *[This chart is similar to Figure 21-3.]* "*Projecting code production gives you another view of what you are up against.*"

The analysts then asked questions and took exception to my answers. I put up more charts. We went round and round while Colonel Roper sat silent. Finally he cleared his throat. "*When can you provide a similar assessment?*" he asked the senior analyst.

"*We will have to collect a lot more data,*" the analyst answered. "*It will take some time.*"

"*This young man seems able to do it while we sit here,*" the Colonel said abruptly. "*I'll give you two weeks.*"

Under his breath the Colonel muttered, "*This smells more like Noran every day.*"

I assumed he had just lived through a similar software disaster and didn't like the prospects of another go around.

The meeting broke up formally. As we were standing around, another analyst said the latest data points had just come in. "They will show a better picture," he insisted. He ran down the hall for them and I put them in the computer. Process productivity had dropped a tenth of an index point—not much, but not a good sign.

Fall 1990

Of course, Shining Shadow was none of my business and I didn't hear anything for more than a year. Then Captain Kramer called to say that funding was available to analyze its status. He did mention that the US Air Force had reassigned the young captain who had been project manager.

We put the last year's worth of actuals into the computer. Our previous projections were within 90 to 93 percent accurate for code production, effort expended, and schedule. Completion might come a month or so earlier than our previous estimate, perhaps by March 1992.

Spring 1991

The Gulf war diverted resources and money was tight. At any rate the government canceled Shining Shadow in May 1991, a year or so from completion. By that time it had overrun to $25 million. The department got some hardware, but no working software.

It was not a happy ending. I felt frustrated even though I had no formal responsibility. Our methods projected what was going to happen quite accurately. Certainly we had raised the issue early enough for

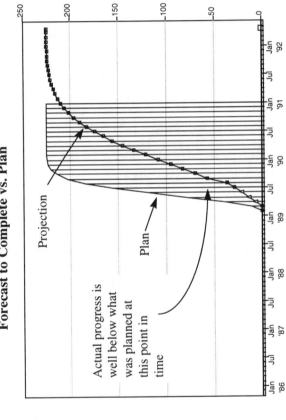

Figure 21-3. The bars represent the existing January 1991 code production plan. At the organization's actual process productivity, the replanned schedule runs to the spring of 1992.

someone to do something. I tell myself that a half hour briefing is not enough time to overturn years of existing administrative practice. Still, I feel sad whenever I think of Shining Shadow. It no longer shines.

Part IV

Reliability

"Quality is Job One," the Ford Motor Company has been proclaiming for more than a decade. In the sense of setting a direction, so be it. In the sense of setting out the work to be done, quality is jobs one, two, and three.

Job One. The design must be right. It must also define a "right" product that is reliable. Building the wrong product, that is, a product that doesn't meet the user's needs, doesn't get you anywhere. Further, building a product that would meet user's requirements—if it operated reliably—doesn't help much. Getting the design right and reliable is the first step to Market-Driven Quality.

Job Two. Manufacturing, inspection, and test operations have to reliably replicate this "right and reliable" design. To the extent that manufacturing (or design) fall short of complete reliability, field service (or recall or upgrades) must fill the breach.

Job Three. The entire process from user needs through design, manufacturing, and field service must be continuously improved. The outcome of process improvement is a still more reliable, less expensive product.

In software Job Two, manufacturing is less important than it is in hardware. Reproducing code on a medium such as a disk is simple and accurate.

Job One, however, utilizes huge amounts of error-prone hand labor. Doing this job better is the topic of this Part. Improving the process, Job Three, is the focus of Part V.

Chapter 22

The Path to Quality

Our process was unacceptable: we were inspecting quality into the product, we were not manufacturing quality into the product. —K. Theodor Krantz, President of Velcro USA on the use of traditional inspection [1]

Once upon a time a solo artisan produced a unique product and sold it directly to the user. He prided himself on its quality and had direct feedback from the user if quality was lacking. Then organizations grew larger and users receded to the far end of a marketing chain. Achieving quality no longer took care of itself.

Inspection

Inspection at the final step in a series of manufacturing operations was an early means of assuring that customers received a satisfactory product. Figure 22-1 shows how inspection fits in the traditional manufacturing and rework process. Inspectors compared characteristics of the product against those prescribed by the engineering drawing or product specification. Units that failed requirements went to rework. If they could be repaired, they went forward as product. Otherwise, they were scrapped.

The final-inspection paradigm has flaws that gradually became apparent. In the words of the Velcro executive whose organization had been following this practice until the mid-1980s, "Quality determination coming at the end of the production process had to be relayed back up the line, and such feedback is often incomplete, unreliable, and certainly untimely. Besides, we were throwing out a certain amount of product because it was defective or didn't meet specifications. We were wasting all the value we had added in the production process, like dyeing, coating, and slitting. Moreover, when we found a problem at the end of the process, it might have run for several days before being spotted" [1].

Traditional Rework

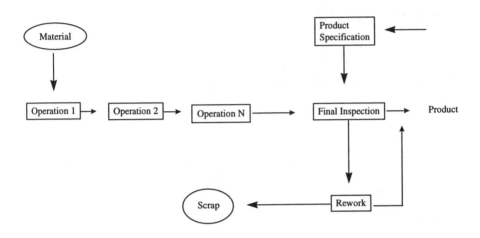

Figure 22-1. Inspection occurs at the end of the traditional manufacturing process. There is an expensive flaw in traditional final inspection: defective products pile up at the end of the manufacturing process until feedback eventually corrects the process deficiency.

In fact, Velcro USA was throwing away five to eight percent of its production, depending on the product line. In the first year after getting more serious about quality, the company trimmed this waste by half. In the second year they saved another 45 percent. They became more competitive not only on quality, but also on price.

Fortunately, in many manufacturing plants ostensibly following the final-inspection paradigm, alert operators or supervisors often noticed a deficient operation. They corrected it before the feedback loop through final inspection had time to do so. We remember walking through a noisy factory with the superintendent. "Uh, oh," he interrupted our conversation, "that doesn't sound right." He rushed over to the other side of the shop and stopped a press. Unfortunately, not all deficiencies announce themselves by a change in the composition of factory sounds.

Of course, progressive factory managers began to institutionalize this concept long ago. Figure 22-2 shows the flow of incremental correction. After each manufacturing operation, they measured the characteristics produced by that operation. If the characteristics were out of specification, they adjusted the operation right away, holding the number of defective units to just a few.

Incremental Correction

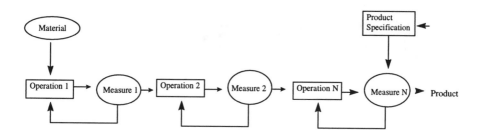

Figure 22-2. By measuring the product after each operation, progressive managers corrected manufacturing deficiencies at once.

Statistical process control

The next logical question is what is "out of specification"? Suppose Operation 2 in Figure 22-2 is producing a machined part with one key dimension. That dimension is usually specified with a tolerance: 2.000 ±0.006. Supposedly any part with a dimension between 1.994 and 2.006 will work satisfactorily in the next higher assembly. Under normal production conditions a sample of the production is measuring right around 2.000— between 1.998 and 2.002. One day, however, the samples begin to measure between 2.004 and 2.006. The part is still within tolerance, but the inspector suspects that the machine is beginning to run out of limits. Something systemic is wrong and needs looking into.

To keep track of what's going on, inspectors or operators plot their measurements over time on a statistical control chart, such as that in Figure 22-3. On this chart the process went above the upper control limit, but was still within the upper tolerance. The operator adjusted the process and for some time it ran between the control limits. Then it rather suddenly plunged below the lower control limit. In spite of efforts to bring the process back, it soon went below the lower control limit. The parts had to be rejected.

Process Measurement

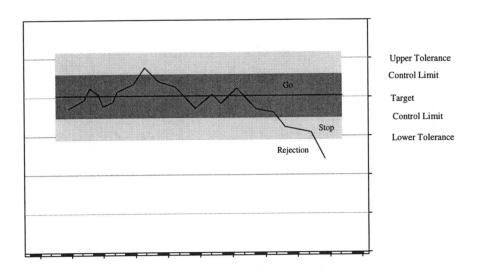

Figure 22-3. An operator plots actual measures over time of some process variable (solid line) in relation to its target value and upper and lower control limits. The limits are within the tolerances permitted.

In any process there are two kinds of deviations from the target dimension: random and systemic. We expect random deviations. Nothing runs perfectly or, for that matter, can be measured perfectly. To the extent that deviations above and below the target dimension follow a random pattern, the process is in control. Random deviations follow the normal distribution, most of them near the target, a few farther away.

Systemic deviations result from something wrong with the process that builds the part—the process is not working normally. Systemic deviations show up on the control chart well away from the target—the average of the measurements is no longer on the target mean. This circumstance is a signal to the operator to adjust the process to bring the measurement average back on target.

The control limits are often set at plus and minus one standard deviation; that is, we expect 68 percent of the measurements to fall within these limits. If the measurements show this pattern, the variations from target are probably random and the process is in control. If successive measurements begin to show some other pattern, the variations may be systemic, and we need to adjust the process.

Beyond inspection to quality

Quality is a broad concept. It is more than just inspection, whether at the end of production or during the process. One definition has it that "Software quality is a set of measurable characteristics that satisfies the buyers, users, and maintainers of a software product" [2]. T. Bowen compiled one set of software quality attributes [3]

Correctness	Portability
Expandability	Reliability
Efficiency	Reusability
Flexibility	Survivability
Integrity	Usability
Interoperability	Verifiability
Maintainability	

Most of these attributes cannot be "inspected in." They must be "designed in." According to Genichi Taguchi, for instance, "the 'robustness' of products is more a function of good design than of online control, however stringent, of manufacturing processes" [4]. This robustness begins with the user. Figure 22-4 shows how customer feedback influences product design. If the users' true needs are not being satisfied,

Using Customer Feedback

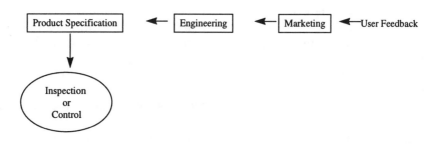

Figure 22-4. An organization works for its customers. Their needs should guide its operations.

they feed their real needs back to the product specification through marketing and engineering. The revised specification leads to a redesign of the product or a revision of the manufacturing process.

The ultimate measure of software quality is customer satisfaction. Satisfied customers must be the goal of enterprises that intend to be continuously successful. "A failure to satisfy a customer, who is everyone from the next person in the process to the end user, is a defect," said Bill Smith of Motorola [5].

Good requirements, good quality

Satisfaction has proved to be a bit hard to measure directly, though sales people and service people in contact with customers have some sense of it. Many companies like to have executives and managers spend some time with customers to gauge this feeling. It is particularly hard to measure satisfaction during software development when customers have yet to see the product. At best, a few representatives of users have participated in the definition of requirements.

Then, if these representatives fully sense the needs of users, we might hope they can communicate them to the software developers. Figure 22-4 symbolizes this link from marketing to engineering. The developers, specifically the requirements analysts, must still get these needs reflected in a specification to guide the rest of the project. An organization that meets the needs of users is customer-oriented. This attitude is a basic tenet of Total Quality Management. Our challenge is to apply TQM to software development.

In the beginning, there were no software products. Users did not know what software could do for them. They found it hard to describe what they did in terms meaningful to software developers. On their side, developers understood the capabilities of computers and programs, but did not fully comprehend users' needs. The outcome was that requirements were often not completely expressed. Thus, because quality from the developers' perspective consists of satisfying the specifications, systems based on inadequate requirements definition were ipso facto of poor quality.

Fortunately, the situation is no longer so bleak. Since the early days of software development, several improvements have appeared. One came about naturally, just through the passage of time. Much present-day software is the end result of a series of product releases. In many fields users have learned what they need through experience with earlier products. Over the years organizations have successively refined requirements for particular applications. Correspondingly, the quality of the system these requirements represent has improved.

This lesson can be applied concurrently wherever users and developers already have experience with software that performs similar functions. Drawing on this experience, they can write better requirements and in this way better meet users' real needs.

In more novel applications, developers cannot replicate in a few months the experience gained over years of product releases. However, they can prototype the parts of a proposed system that are totally new. They can code a new algorithm, a new input or output screen, or a new procedure, and let representative users try it out. On the basis of this experience, what was ill-defined in the requirements can be firmed up. Again, quality is better defined.

After the initial specification, as high-level design proceeds, both developers and customers often gain a better appreciation of the application. They would like to reflect this appreciation in the specification. Remember, a legitimate modification of the requirements leads to an improvement in customer-perceived quality. So, requests for change cannot be lightly dismissed. What to do?

Well, making this decision is a balancing act. The effect of the change on product quality must be balanced against its impact on the length of the development schedule and the amount of effort, number of people, and cost. However, if a change is not fundamental to the product's initial functioning, those concerned may decide to bundle it and other desirable but not crucial changes for a later release.

The point is that good requirements lead to good quality. Total Quality Management principles dictate that you put a good deal of emphasis on this initial phase of software development.

Lessons learned

The experience of other disciplines with quality leads to a few essential lessons:

- ❑ Customers' needs are the starting point for improving the product and the production process.
- ❑ Inspection only at the end of a series of operations leads to waste of material and production time.
- ❑ Measurement and readjustment immediately after each operation correct defect-causing conditions before more material and time are lost.
- ❑ Bringing every worker into the quality circle extends the reach of quality efforts.

These principles can be applied to the software development process.

References

[1] K.T. Krantz, "How Velcro Got Hooked on Quality," *Harvard Business Rev.*, Sept.–Oct. 1989, pp. 34–40.

[2] J.A. Clapp and S.F. Stanton, "A Guide to Total Software Quality Control," RL-TR-92-316, Vol. 1, Rome Laboratory, Air Force Material Command, Griffiss Air Force Base, New York, Dec. 1992, 100 pp.

[3] T.P. Bowen, J.T. Tsai, and G.H. Wigle, "Specification of Software Quality Attributes—Software Quality Evaluation Guidebook," AD A153-990, RADC-TR-85-37, Vol. I, II, III, Rome Air Development Center, Air Force Material Command, Griffiss Air Force Base, New York, 1985.

[4] G. Taguchi and D. Clausing, "Robust Quality," *Harvard Business Rev.*, Jan.–Feb. 1990, pp. 65–75.

[5] B. Smith, "Making War on Defects," *IEEE Spectrum*, Sept. 1993, pp. 43–47.

Chapter 23

A Measure of Software Quality

You'll always get the good news; it's how quickly you get the bad news that counts. —Harvey Mackay, author of *Swim with the Sharks: Without Being Eaten Alive* [1]

Mechanical parts must measure within a tolerance of a specified dimension if they are to assemble successfully with other parts. If a dimension, as manufactured, is not within tolerance, the part has failed. It may look handsome and it may have been machined by a worker of fine character, but it is defective.

Measures of all pertinent dimensions ensures that the entire product can be assembled. The completed assembly then meets the first test of quality: it exists as specified. Whether it meets users' needs is a function of the feedback from users to the design and manufacturing process.

The specified dimensions are themselves the outcome of a long process of requirements determination, specifications writing, drawings preparation and checking, tolerance analysis, model building, testing, and so on. If engineers do not carry out this process skillfully, the dimension on the drawing itself may be incorrect. The assembly does not go together.

Thus, hardware engineers face a twofold task. The design process must result in correct dimensions. The manufacturing process must reproduce those dimensions consistently.

A software measure

Software developers need a more immediate measure than ultimate customer satisfaction—one comparable to the dimension in hardware. The design process is much the same: requirements, specifications, design, coding (analogous to drawing), and testing (analogous to model testing).

The software manufacturing process reproduces the code on a medium that can be distributed, such as disk. Manufacturing in this sense is so simple and reproduces the code so accurately that we need give it little thought, unlike the manufacture of physical parts.

Unfortunately, if software engineers commit an error anywhere in the development process, it passes through to the code and thereafter is unerringly reproduced. It is evident, then, that attaining quality in software boils down to obtaining a correct reflection of users' needs in requirements and specifications, and then accurately translating these requirements into design and code. Going from users' needs to requirements is to some extent intuitive and perhaps nonquantifiable. Nevertheless, the requirements are stated in writing, or should be. Anything in writing can be pored over to remove errors. Similarly, design, code, test plans, and any other work products of the software development process can be scoured to remove errors.

Thus, in software the "dimension" to measure is errors during the development process. Does each item in the requirements accurately set forth a user need? Do the specifications correctly reflect the requirements? Does each element of the design—algorithms, data structures, control flow, and so on—carry out a specified item? Does the code accurately implement the design?

At the end of each work step in the software development process, developers should "measure" the dimension, that is, review or inspect their work-product and count the defects. Soon after, they correct the defects and, if necessary, reach back to correct previous work products from which the current product derives.

If an organization counts the errors that turn up at each stage of development, that count is then an indication of quality and more specifically a measure of reliability. If the organization has a record of error counts on previous projects, it has a basis for judging what the current error count means. In Figure 23-1, for instance, the current project is finding fewer defects than the average of past projects. This comparison suggests that the current project is going well.

In reality, of course, development stages overlap. Errors are not always discovered during the stage in which developers commit them. Therefore, it is more practical to control against an error-rate curve, a projection of defects per month, as drawn in Figure 23-2. In effect, the record of the past defect rate is a first approximation of the defect rate on the next project. Given this approximation, project leaders can track current errors against it.

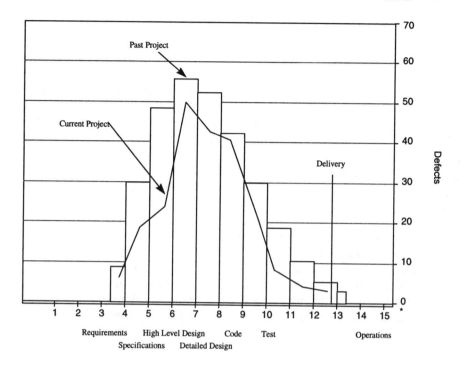

Figure 23-1. Developers can compare a count of errors in each development stage with the average count of previous similar projects.

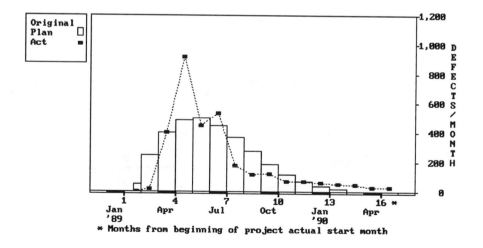

Figure 23-2. The bars represent defects per month on comparable past projects and the dotted line, the defects discovered each month on the current project.

Projecting the software measure

Software development needs something comparable to the statistical control chart described in the last chapter. In manufacturing, this chart projects a target and upper and lower limits against which to assess the measurement of dimensions. In software, a curve that projects the expected number of defects per month, supplemented by upper and lower bounds, would serve this purpose, as Figure 23-3 shows. Each month the actual count of defects could be compared with the projection.

In developing manufacturing control charts, manufacturing people have continuing experience to draw on. In software development each project is new, so deriving control charts from project experience is more difficult. We can start with the thought that errors are proportional to the work done, the work done is proportional to the effort put in, and the effort put in follows a Rayleigh curve. Figure 23-4 shows this pattern.

The application of people follows the first curve. The area under the curve is proportional to effort. This curve begins with detailed logic design and continues through coding, test, and operation and maintenance. Prior to the main build, two preliminary phases, feasibility study and functional design, not drawn on this figure, have taken place.

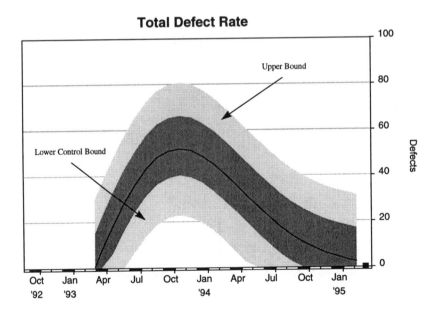

Figure 23-3. Defect rate against time, bounded by upper and lower limits, can serve the same purpose in software development as the statistical control chart in manufacturing.

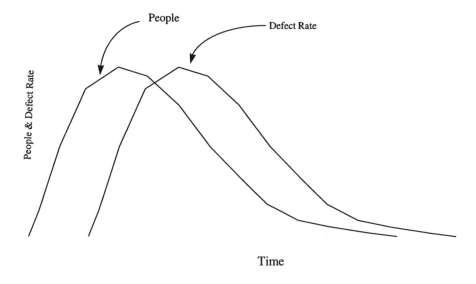

Figure 23-4. In general the defects (detected) rate curve is similar to the people curve.

The second curve is the overall defect rate, not just defects in code, but defects in all the work products of software development. Defects may be injected during requirements analysis, specification writing, functional design, and detailed design. If they are not immediately found (by inspections, reviews, or walkthroughs) and corrected, they turn up later in code. Not until a defect is present in the code can a module or system test find it.

The defect-rate curve is similar to the people curve, but lags it by the time required to find the defects. This time lag varies with different development practices. If an organization employs reviews, inspections, and walkthroughs all during the development process, most defects are discovered shortly after they are committed. Thus, the defect-rate curve closely follows the people curve in the figure.

Code production gets under way after the detailed design phase, builds up to a peak during coding, and then declines during test. If an organization depends primarily on testing to find defects, defects are necessarily discovered quite late in development—after code has been integrated and test becomes possible. Figure 23-5 shows this pattern.

Software organizations still in the final-inspection era of quality control (inspection is done only at the very end of the process) devote little organized effort to detecting and correcting defects until the program gets to system-integration test. Then, defects in the code being plentiful, they find many. As testing proceeds, they soon detect all the easy-to-find

Rayleigh Profiles (Defects Detected During Test)

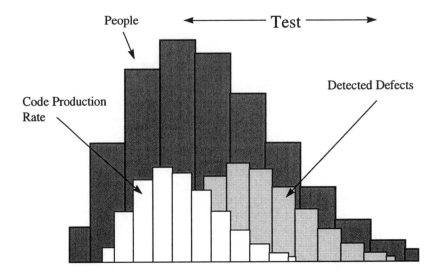

Figure 23-5. Using a Rayleigh curve to forecast the defect rate. The Rayleigh curve shows the up-and-down nature of the defect rate during the entire development period.

defects. Additional defects become harder to find and the rate of detecting defects drops off along an exponential curve.

Many investigators have based defect projections on this exponential decline. Theoretically this type of curve begins at a peak, declines very rapidly at first, then more slowly, finally becoming asymptotic to zero. That is as it should be. There can be no guarantee that testers have eliminated every last defect from any software program. Some unusual set of circumstances that has not appeared before may reveal one more defect [2].

Rate forecasting

From defect data recorded by many companies, we have worked out the mathematical relationships that enable a software organization to forecast the defect rate it will experience on a coming project. Statistically, it is the expected rate, or the mean rate. By the same relationships we also forecast upper and lower bounds, set at plus and minus some number of standard deviations from the mean rate. With the help of this forecast, comparable to a statistical control chart, you are in a position to track your errors.

A project that is experiencing 30 defects per calendar month is finding an average of one defect per day; that is, its mean time to defect—the reciprocal of the error rate—is one day. Thirty defects per month is equivalent to 1/30 month to defect.

Note that we express MTTD in units of calendar time, such as month or day. When we integrate a software system and begin to operate it, in test or operations, the system then experiences failures. We then have a mean time to failure. "Time" in MTTF is usually execution time or operational time, not calendar time.

We find that managers like to see the defect rate forecast with the Rayleigh curve. The portion of the Rayleigh curve that occurs during test is similar to the exponential curve. The entire length of the Rayleigh curve has the added advantage of showing the up-and-down nature of the defect rate during the entire development period. The curve asserts that developers are committing errors during requirements, specifications, functional design, and detailed design, as well as coding. The curve reminds you of the wisdom of employing inspections, reviews, and walkthroughs to find these defects and, of course, to fix them as work goes along.

References

[1] H. Mackay, *Swim With The Sharks: Without Being Eaten Alive*, William Morrow and Company, New York, N.Y., 1988, 313 pp.

[2] S.L. Pfleeger, "Measuring Software Reliability," *IEEE Spectrum*, Aug. 1992, pp. 56–60.

Chapter 24

Plan Project to Minimize Defects

The cost of continuing to develop failure-laden software with its associated low productivity can at best increase cost and at worst so affect an organization's competitive position that it is difficult to remain in business. —Richard H. Cobb and Harlan D. Mills [1]

The obvious course is to reduce defects. Not so obvious is just how to do that. At first blush it seems that reducing defects is a technical effort in the province of software technologists. Indeed it is, at least in part, but the way you plan to do a project has a substantial effect on the number of defects that result [2].

The five management numbers—development time, effort (or cost or staff), size, process productivity, and manpower buildup—have an effect on the number of defects incurred on a project. Consequently, planning these management numbers in appropriate ways can reduce the number of defects. Moreover, all five factors are interrelated. That is, in planning a project, when you change one, it affects the others. In this chapter we show how each affects defects separately, but keep in mind that when one moves the others move, too.

Development time: more is better

On the one hand, if the development schedule allows adequate time for people to think carefully and check their own work every few hours (and immediately correct their own errors), we expect they would make fewer countable errors. On the other hand, if the schedule is tight, if developers seize on the first idea that comes to mind, if they have no time for self-checking, we expect them to make more errors.

"If you're under enormous pressure to deliver a stop-gap system, then of course quality is likely to suffer," Tom DeMarco observed many years ago [3]. In fact, our database now unequivocally establishes that quality will suffer when you try to crowd the schedule. Figure 24-1 shows that the number of defects declines very substantially as planned development time increases by up to 35 percent.

The minimum time shown is the time below which no comparable project has ever been completed. At this minimum development time, the number of defects is at a maximum. Beyond 135 percent (1.35) the curve declines slowly. The number of defects falls only slightly. The implication is that, once the planned development time is sufficient, people make about one fourth to one third as many errors as they make when rushed. Time beyond this sufficiency level doesn't seem to reduce the error rate very much more.

The figure shows that it is best to *avoid the temptation to develop a system in minimum time*. If you do, you reap the maximum number of errors. It also shows that you suffer fewer defects if you *allow as much planned development time (up to 135 percent of minimum) as your constraints permit*.

Ratio of Defects to Defects at Minimum Time

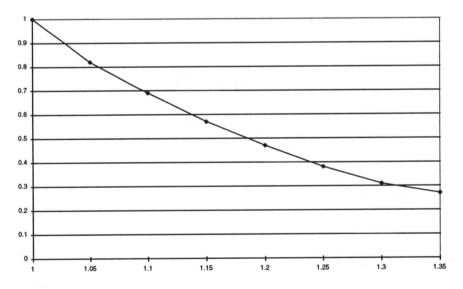

Figure 24-1. Both axes are expressed in ratios, but the lesson is clear: the number of defects declines if planners allow a little more than the minimum possible development time (1/1).

Effort

Effort, staff, and cost are roughly synonymous. Effort occupies the area under the Rayleigh curve for staffing. Multiplied by a salary-rate-plus-overhead factor, it is approximately equal to project cost. The top of the people curve is peak staff and across its middle we could draw the average manpower line. A number of key software process variables move in harness with effort. Therefore, when we observe that the number of defects increases as effort grows, the observation applies also to staff and cost.

What happens when the number of tasks exceeds the capacity of one person and the tasks must be divided among several people? Well, the time devoted to communications among the people must be added to the time taken by their separate work. For two or three coworkers this communications time is not great. To the contrary, there is often pleasure and profit in working together. The story is different when large numbers of people must communicate with each other. As Figure 24-2 reveals, communication time increases exponentially. The number of communication relationships is given by $n(n - 1)/2$, where n is the number of people involved.

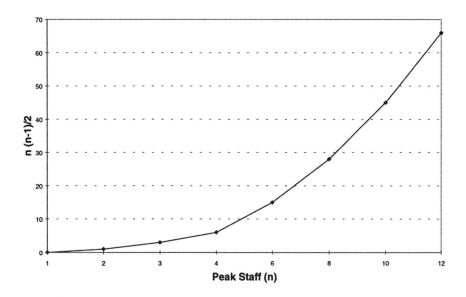

Figure 24-2. The increase in communication interactivity as project staff expands is paralleled by the increase in the number of defects, as Figure 24-3 shows. The number of communications relationships is given by n(n − 1)/2, where n is the number of people involved.

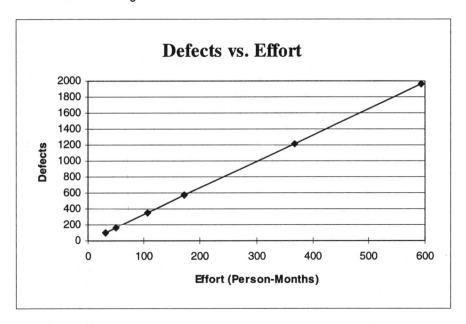

Figure 24-3. The total number of defects incurred on a project increases with effort. This diagram is based on a 100,000 SLOC business system accomplished at process productivity indexes of 12, 16, or 20 and manpower buildup indexes of 1.8 or 3.0.

"Since software construction is inherently a systems effort—an exercise in complex relationships—communications effort is great, and it quickly dominates the decrease in individual task time brought about by partitioning," Fred Brooks advised us a generation ago [4].

To the communication time among workers, we might add the time spent going up and down and across hierarchies of managers. We might add the misapprehensions—often leading to errors—resulting from poor communication. We might refer you to the hundreds of amusing incidents of confusion in the software village, as observed by Gerald M. Weinberg and his anthropologist partner, Dani Weinberg [5]. But you live in the village, too. Instead, we merely draw Figure 24-3 showing that in our database an increase in the number of defects goes along with an increase in effort (and peak staff). We speak in numbers!

To draw this defect line, we projected management numbers for the same 100,000 SLOC business system at six combinations of process productivity and manpower buildup:

Productivity Index	Manpower Buildup Index
20	1.8
20	3.0
16	1.8
16	3.0
12	1.8
12	3.0

The average productivity index for business systems was 16 in 1992 with a standard deviation of ±4 index points. Thus, Figure 24-3 represents the productivity range within which most business systems were being accomplished. Similarly a manpower buildup index of 3.0 was about average, and 1.8 was a little slower than average.

As we have previously noted, the management numbers are interrelated. So when we improve process productivity we shorten development time; when we slow down manpower buildup we lengthen development time. Figure 24-3 depicts the effect of increasing effort on the total number of defects, but that changing effort is itself the result of changing process productivity and manpower buildup. Nevertheless, the general picture is clear: When we put the horde of Attila the Hun on a project, they make a lot of mistakes.

The figure shows that it is wise to reduce the amount of effort (staff or cost)—not arbitrarily but thoughtfully, by extending development time, improving process productivity, or slowing manpower buildup. More calendar time with fewer people at less pressure reduces defects.

Size

The total number of defects increases with the size of the project, as we saw in Chapter 4 (Figure 4-3). The deviation of the number of defects from the mean is very great. The numbers in projects of the same size differ by orders of magnitude. Consequently, size alone appears not to predict with much precision the number of defects to be expected.

This tells us that reducing the size of the proposed system will proportionately decrease the number of defects. You can reduce size sometimes by resisting bells and whistles (and marginal functionality) and sometimes by deferring parts of the system to a later release.

Process productivity and manpower buildup

As process productivity improves, the number of defects declines. Figure 24-4 plots defects per thousand source lines of code against productivity indexes of 12 to 20 for the same 100,000 SLOC business system used in Figure 24-3. The two curves, MBI = 3.0 and MBI = 1.8, indicate that a

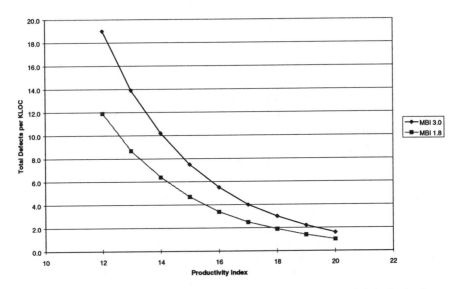

Figure 24-4. As process productivity improves, the number of total defects declines. Slower manpower buildup results in fewer defects.

project performed at a more rapid manpower buildup incurs a greater number of defects.

The figure suggests two action items to reduce defects. First, you should improve your process productivity. (Part V deals with process improvement.) Second, you should build up staff at a reasonable rate, certainly no more rapidly than work units have been broken out to engage additional people.

Delivery date

You must establish the value of all five factors—development time, effort (or cost or staff), size, process productivity, and manpower buildup—at the time you plan the project. Then you must follow the plan. Of course, you may have set unrealistic factors. You may now be coming to the end of the development time you set and find that the remaining defects are still too high for the nature of the application or for your customer's desires. At this point about all you can do is defer delivery beyond the planned date and use the additional time to continue testing and fixing defects.

Extending the delivery date has the advantage of reducing the defects remaining at delivery. The user can operate for longer periods between failures. The user spends less time waiting for failures to be repaired. Of course, extending the delivery date has drawbacks. The customer does not get the benefit of the system during the delivery delay. The vendor lays out additional effort and cost and perhaps is using people who were planned to staff later projects.

For example, we calibrated our curve of expected defects to reach 95-percent reliability at the same point as the milestone named full operational capability. Thus, under this calibration arrangement five percent of the total defects remain in the code when it first becomes operational. These remaining defects turn up and get fixed during the operational and maintenance phase—the long tail of the Rayleigh curve.

There is a distinction between "planned development time" and "extended delivery date." The first must be set at decision time or planning time—also the time for setting tradeoffs. The "extended delivery date" may be set at any time that management determines the intended application requires greater than 95-percent reliability.

In many applications 95 percent reliability is adequate. For applications in need of greater reliability, you can project the points at which from 95.0 to 99.9 percent of the defects have been detected. The time to remove the last few percent of the defects is usually a matter of months. Our experience is that only users needing high reliability wish to expend this additional time and effort.[1]

Thus, when you need better reliability or a longer meantime to defect, you should extend the delivery date.

If the software doesn't have to work, you can always meet any other requirement. —Gerald M. Weinberg's "Zeroth Law of Software Development" [6]

References

[1] R.H. Cobb and H.D. Mills, "Engineering Software under Statistical Quality Control," *IEEE Software*, Nov. 1990, pp. 44–54.

[2] L.H. Putnam and W. Myers, "Haste Makes Waste: The Way You Plan A Project Affects Its Reliability," *American Programmer*, Feb. 1992, pp. 21–27.

[3] T. DeMarco, *Controlling Software Projects*, Yourdon Inc., New York, N.Y., 1982. 284 pp.

---

[1] A rule of thumb derived from our data states: it takes 1.25 times the nominal development time (that is, the period to full operational capability) to achieve 99.0 percent reliability; and 1.5 times longer than the nominal development time to realize 99.9 percent reliability.

[4] F.P. Brooks Jr., *The Mythical Man-Month: Essays on Software Engineering,* Addison-Wesley Publishing Co., Reading, Mass., 1974, 195 pp.

[5] G.M. Weinberg, *Quality Software Management, Volume 2, First-Order Measurement,* Dorset House Publishing, New York, N.Y., 1993, 346 pp.

[6] G.M. Weinberg, *Quality Software Management: Volume 1, Systems Thinking,* Dorset House Publishing, New York, N.Y., 1992, 318 pp.

Chapter 25

Institute Early Inspection

Successful management of any process requires planning, measurement, and control. In programming development, these requirements translate into defining the programming process in terms of a series of operations, each operation having its own exit criteria. Next there must be some means of measuring completeness of the product at any point of its development by inspections or testing. And finally, the measured data must be used for controlling the process. —Michael E. Fagan [1]

The way you plan to do a project can minimize the number of errors committed, but analysts, designers, programmers, and coders will still fall short of perfection. You can expect some considerable number of errors to be made at every step of the software development process.

For example, Tom Gilb and Dorothy Graham expect to find 10 to 20 "major defects" per page of work product prior to "Inspection" for developers who are not accustomed to having their work inspected [2].*

As another example, Edward F. Weller reported 5,649 major defects removed from 6,870 design-document pages inspected during 2,823 code inspections and 348 document inspections in 1992 at Bull HN Information Systems. The work inspected was an operating system totaling more than 11 million SLOC to which 400,000 to 600,000 SLOC are added each year [3].

At the code level, the number of errors or defects is also high, as we showed in Chapter 4. Figure 4-3 plotted the number of errors between system integration testing and full operational capability. The number is from one error for systems of a few thousand lines of source code up to 10,000 for a one-million line project. Because only about 17 percent of the

---

* A Gilb-Graham inspection finds issues that may become defects after investigation. "Major" is a severity classification that identifies a defect expected to significantly increase cost if not found until later in the development process, such as in test or use. "The primary purpose of this classification," they say, "is to help ensure that the inspection process is directed towards economically useful work."

defects are usually found between system integration testing and full operational capability, the total number of defects over the entire development period would be from about 6 to 60,000.

It is evident, then, that software people have always devoted much energy and skill to finding and fixing errors, since the number of defects at the time of release is only a small fraction of the number committed. Even so, our experience is that much software is being delivered at about the 95 percent reliability level—about 95 percent of the total defects have been removed. That level may have been passable once upon a time, but as systems grow larger and operate in more critical realms, it is no longer satisfactory. Software people must learn to do better. Many are doing better, showing that higher reliability is within the state of the art. Reliability levels better than 99 percent are being achieved.

Testing

The traditional defense against defects has been to find them and fix them. Integration and system testing are the most common "find 'em" techniques. They are analogous to final inspection in the manufacturing process and suffer from the same gross deficiency. They don't find defects along the way. You can't test until you have code, but by then many errors have crept in during requirements analysis, specification writing, and design. An error early in the development process tends to multiply into several defects by the time it gets into code.

When testing finds a failure, it is usually not obvious in which line of code the fault lies. Testing found only the effect, not the cause. Many person-hours may be spent isolating the fault to the line of code that can be fixed. Moreover, considerable calendar time may pass before the programmers find the line, putting schedules in jeopardy. The fix itself, especially if made by those other than the original developers, may propagate further defects throughout the code.

If the defect remains hidden until the software becomes operational, the cost may be even more severe. The international funds-transfer system may be down while interest costs mount. Worse, an aircraft may crash. A software-guided missile may impact at the wrong time and place.

With these drawbacks, why is testing so popular? One reason, of course, is that it is the last line of defense before release to the user. Moreover, most customers insist on some testing and may even do some themselves.

Another reason is that testing is automated in various degrees. Test programs do most of it. In contrast, upstream methods of seeking defects—which take place before operating code is available—depend heavily on human effort, a resource always in short supply.

Upstream efforts

The less common "find 'em" techniques—reviews, walkthroughs, and inspections—have in common that they take place upstream of testing. They differ in their degree of formality—from relaxed to rigorous.

Design reviews

At the relaxed end of the design-review scale is the irregular scrutiny of the emerging design by the designer and close colleagues. Older engineering disciplines long ago formalized this scrutiny into periodic technical reviews by appointed groups of concerned persons. This practice has been carried over into software engineering by the US Department of Defense and others with specified reviews at the conclusion of the feasibility study, functional design, and detailed design. At a minimum the design review uncovers blind spots, biases, or mistaken notions of the original designer. At its best, with clear goals and procedures and time to do the work, it can greatly improve the initial design.

Walkthroughs

A test program checks many paths through the code very rapidly. In a walkthrough several designers or programmers trudge slowly through the design or code logic, step by step, hence the name. For large programs, people may not have time to think through all the possibilities. They should find it worthwhile to walk through the frequently used ones.

When a program can find errors, it is only sensible to use it. A syntax checker does its work faster and more accurately than we can. Unfortunately there are not checking programs for every kind of defect, particularly not for those due to flaws in thinking through a sequence logically or, even worse, those for which a logical step is completely missing.

Inspections

"Inspections are a *formal, efficient, and economical* method of finding errors in design and code," Michael E. Fagan observed in 1976 after several years of applying what he called "inspection" on good-sized projects [1]. "Inspection" is a widely used word, but Fagan meant something very specific by it. He summarized the Fagan inspection method in five stages:

1. Define a series of operations making up the software process, each of which has exit criteria that can be inspected.
2. As each inspection team runs through a series of tasks extending from overview to rework, keep the team focused on one objective at a time, such as finding errors or fixing errors.

3. Classify errors by type and frequency as a guide to further inspection activity.
4. Teach inspectors how to look for each type.
5. Use inspection results to improve the software process.

Inspection lessons: early and often is best

"Vote early and often" was the cry of the big-city machine bosses in now bygone days (we think they are bygone). Software managers might well adopt a similar slogan: "Inspect early and often."

One of the projects Weller described was not inspected early; it first used inspection at the coding stage. Requirements and design documents had not been inspected. Project members inspected more than 12,000 lines of C and found an average defect density of 23 defects per thousand source lines of code. The team felt they had found a respectable number of defects, but the team leader noted that they had assumed a correct design. A postmortem review of the project showed that the defects the team found were mostly coding or low-level design errors [3].

In other words, code inspectors, assuming correct design, had not found deficiencies in requirements or defects in high-level design. The lesson is "inspect early"—deficiencies should be hunted down and removed at each stage of the development process.

We can imagine a series of Rayleigh curves representing rates of effort, defects created, defects discovered, and defects corrected. The rate at which defects are created would follow the effort curve closely in time. Detection might occur soon after creation—a few days or a few weeks—if the software development organization employs frequent reviews and inspections. Detection might be delayed for months, however, if the organization waits until the code is ready to operate in test before it seriously looks for defects. Defects are then corrected some time after detection.

A metric used by some organizations is the calendar time between finding and fixing defects. When defects are found in test, this time is often long because what test finds is a failure in operation, not the defect in requirements, specifications, design, or code that led to the failure. Moreover, finding this defect and fixing it is often the responsibility of a person other than the original designer or programmer, so about one in six fixes is itself an error.

When defects are found by an inspection near in time to the original work, the developer is still right there. What he designed is still fresh in his mind. Normally he is able to correct the defect within hours of the inspection meeting.

Of course, project members do not detect all defects by these inspections and design reviews. Some defects slip by until test. However, fewer defects are left in the code at the time system test begins than in the method that depends largely on test to find defects. For example, Fagan estimated that organizations employing Fagan inspections find 60 to 90 percent of defects before testing. The exact percentage depends on the skill with which an organization implements the method.

In addition, developers learn from the inspection process itself. "The programmer finds out what error types he is most prone to make," Fagan continued. "This feedback takes place within a few days of writing the program [enabling him to show improvement on later work]."

This learning leads not only to fewer errors but also to improved productivity. Fagan found that organizations using inspections increased productivity by 10 to 25 percent. On a project at a life insurance company, for example, he cited a productivity increase of 25 percent following the introduction of formal inspections. In this case the productivity increase was measured as a reduction of 25 percent from projected person-days to actual person-days [4].

"An improvement in productivity is the most immediate effect of purging errors from the product by [early] inspections," Fagan said. "Rework done at these levels is 10 to 100 times less expensive than if it is done in the last half of the process."

Gilb and Graham put much the same thought this way: "Each major problem found at Inspection will save you about nine hours of down-stream correction effort, should it show up in either the testing stage or in the field." Referring to defects turning up in the field, they observed, "Inspection can be expected to reduce total system maintenance costs dramatically (at least 10 to one)."

Gilb and Graham summarized typical results from organizations that have introduced inspections:

- ❑ net productivity increases of 30 to 100 percent;
- ❑ net time-scale reductions of 10 to 30 percent;
- ❑ reduction in test execution costs and time scales of five to 10 times, since there are fewer defects to find;
- ❑ reduction in maintenance costs (2/3 due to defect elimination) by one order of magnitude;
- ❑ near "automatic" improvement in software engineering work quality and consequent work product quality; and
- ❑ early or on-time delivery of systems.

"Inspections increase productivity and improve final program quality," Fagan concluded. "Furthermore, improvements in process control and project management are enabled by inspections."

Contrarians

The facts adduced by Fagan, Gilb, Graham, and many others seem quite convincing. For instance, Watts S. Humphrey of the Software Engineering Institute provides the rule of thumb: "It takes about one to four working hours to find and fix a bug through inspections and about 15 to 20 working hours to find and fix a bug in function or system test" [5].

Nevertheless, many organizations are still "quite contrary." Many managers have had experience with great ideas that somehow did not work out successfully. Early inspection is a great idea, but it doesn't work out automatically. You have to work hard to make it work out, but it does work.

Watts Humphrey observes in the foreword to the Gilb-Graham book: "The inspection method is not intuitively obvious. It takes upfront investment, it is often hard to sell to a skeptical organization, and it requires skill."

We hope its merits are now more obvious, but it does require transferring people from testing and fixing late in the software process (and the funds to pay them) to inspection early in the process. It does require taking time and funds to train moderators and participants. It does take time away from urgent work on design and coding. If not done skillfully, it may lead to bad feeling between developers and other members of inspection groups.

If managers do not resist the urge, they may apply information gained from inspections to individual merit reviews. That is a no-no. Inspections have to deal with the work, not personalities. Otherwise developers begin to figure out how to beat the system. We want to *find* defects, not hide them.

In short, like everything else in software development, making effective use of early inspection is hard. It takes organizational discipline. In return, "the eventual savings and benefits will outweigh the costs incurred, including not only the ongoing running costs, but also the startup costs," Gilb and Graham insist.

Boiled down, what these people of long experience are saying is "Do early inspections. They pay."

In our terms, your productivity index goes up. You save development time, effort, and money. Your product is better. Your customers will be delighted. (pleased, anyway).

References

[1] M.E. Fagan, "Design and Code Inspections to Reduce Errors in Program Development," *IBM Systems J.*, Vol. 12, No. 3, 1976, pp. 219–248.

[2] T. Gilb and D. Graham, *Software Inspections*, Addison-Wesley Publishing Co., Reading, Mass. 1993, 471 pp.

[3] E.F. Weller, "Lessons from Three Years of Inspection Data," *IEEE Software*, Sept. 1993, pp. 38–45.

[4] W. Myers, "Build Defect-Free Software, Fagan Urges," *Computer*, Aug. 1990, pp. 112–113.

[5] W.S. Humphrey, *Managing the Software Process*, Addison-Wesley Publishing Co., Reading, Mass., 1989, 494 pp.

Chapter 26

Achieving Quality

Debugging is '99 percent complete' most of the time. —Fred Brooks [1]

Early in test, when the test group is finding scores or hundreds of defects each day, everyone realizes that defect detection and fixing are not nearly complete. Later, when the test group finds defects in onesies and twosies, project people like to think that debugging is "99 percent complete."

Those working to a defect-rate curve, whether declining exponential or trailing Rayleigh, recognize that either curve has a long tail. Defects can be scarce for a long time but the curve implies that some are still there. In fact, the curves extends out into the operational period.

Control against projections

Dynamic tracking and control of defects follow the basic principles we outlined in Chapter 19. The defect-rate detection chart (Figure 26-1) projects defects per month together with control bounds. This chart terminates at full operational capability (end of main build), where about two or three defects per month are still being detected. The tail actually extends out into the operational period.

After the main build gets under way, you can post the actual defects per month, as Figure 26-2 shows. In this case people did little inspection during the detailed design phase. Not until unit test began did a count of defects first appear. Possibly because most defects were left for test to detect, the project revised the defect rate upward and extended the schedule (white squares).

Total Defect Rate

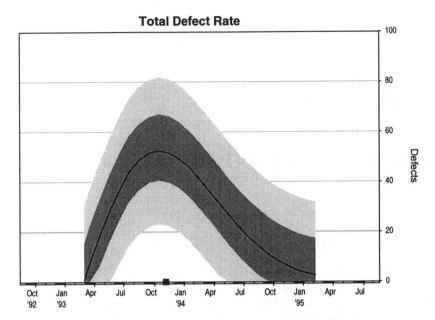

Figure 26-1. The solid line represents the expected value of the defect rate each month during the main build. The upper and lower lines are the limits. Here, the chart ends at the point of delivery, but the Rayleigh curve, representing defects remaining after delivery, would continue out into the operational period.

Total Defect Rate

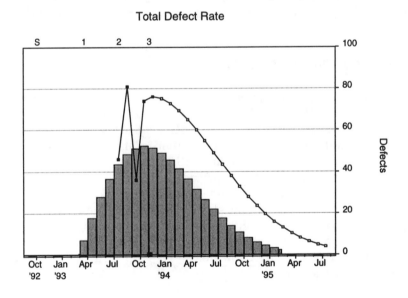

Figure 26-2. The bars represent the original plan; the white squares, the revised projection; and the black squares, the defects actually counted each month.

Prevent defects

So far, we have been talking mainly about monitoring and finding errors. Although we have shown that allowing a reasonable amount of development time leads to fewer errors, why not take direct action to avoid those errors in the first place?

Well, that thought has crossed the minds of more than a few people. The real world—in which software is to perform—is messy and informal. The computer hardware—in which software operates—is precise and formal. The trick is to go from the informal to the formal without committing errors.

Better Requirements

The place to start is requirements—the first step from the real world. If they can be made more precise, developers have a better chance of turning them into accurate code.

Getting something done in the real world is the province of people. In the software field, what is to be done is called "requirements." They are based on information provided by people, whether users or customers. Because users generally have almost endless desires, customer and software development managers are the ones to introduce economic and time constraints within which the requirements must be held.

Users typically have not thought through all the implications of what they initially think they want. So it takes time for the users, requirements analysts, and economic constrainers to hammer out the first take of the requirements.

The software manager is in a position to allow this time. A little more time up front to get the requirements right ensures that the system ultimately produced is the right one. Nothing is more expensive than producing the wrong system. Good requirements make size estimating more precise and increase the probability of completing the project within cost, schedule, and reliability limits.

Despite whatever time and attention users and developers give to requirements in the beginning, they often become aware, as work proceeds, of additional features to add to the initial set of requirements. Many of these add-ons may be entirely reasonable taken by themselves. In spite of their desirability, however, they affect the management numbers.

In proceeding from requirements to specifications and then to design, specifiers and designers may find inconsistencies that need correction. Most of these, too, will appear to be quite necessary. Again, correcting the requirements affects the management numbers.

In both these circumstances, managers are the ultimate constraint enforcers. To help them perform this function, they need a procedure for changing requirements. They need an organized way to bring all the functions affected by a proposed change into the making of the decision to adopt the change. They must ensure that the people representing these functions have thoughtfully considered the effects on staffing, cost, schedule, and reliability.

If analysts can turn the informal requirements document into a formal specification, as we describe below, they can demonstrate that the operations underlying the specification are correct. However, they might still commit errors in the first two steps: (1) extracting the informal requirements from the real-world environment and (2) transforming them into formal specifications. Verifying the correctness of the formally stated specification often highlights errors in both these steps.

Better specifications

Software developers generally write requirements in natural language, such as English. This language, of course, has the great advantage of being understandable by business-side people, but it also has two big disadvantages. First, it is prolix, often resulting in hundreds of pages of requirements. This amount of reading is usually beyond the time availability of most people concerned. Second, it is ambiguous. Workers later in the process often have difficulty establishing unequivocally what the original writers intended.

At the real-world beginning of the software development cycle, what the users want is imprecise and ambiguous. At the computer end, what developers put into the computer—code—is highly precise. The computer does exactly what the program instructs it to do.

Somewhere between the hard-to-perceive world and the precise machine, the statement of what users want has to become highly formal. At the latest this conversion has to take place at the point of writing code. Up to that point, specifications and design methods can be, and often are, indistinct.

Some people want to see this conversion to the formal take place as early as possible. That would be between the requirements stage and the specifications stage. The requirements would be informally stated in a natural language. The specifications would be formally expressed in "a formal language."

The English language is notoriously ambiguous. Words have many meanings, depending on the context in which they occur. In a formal language words have only one meaning. A specification expressed in such a language, therefore, is unambiguous. The process of writing a formal specification reveals any ambiguity, incompleteness, or inconsistency in the informally expressed requirements.

Once system requirements are expressed as formal specifications, they can be processed more accurately through the subsequent stages of software development—high-level design, detailed design, coding, and so on. Developers make fewer errors. The delivered system has fewer defects. The mean time to failure is longer.

Developers have not used formal specification methods widely so far. Some methods have been successful in safety-critical applications, but developers need considerable training to use these methods. They're not exactly something you start using the next month, although one day they will greatly increase productivity and quality.

Then there is the long chain from the formal specification through design, coding, and testing to the operating code. If developers can refine this chain into many simple steps, they can verify each step. For instance, researchers have reduced some of these steps to mathematical formulations subject to proof.

As we discussed in the last chapter, with early inspection at the output of each software stage developers can correct defects at much less cost than if they found them at a late stage, such as test or operations. Similarly, if developers accomplish a logical proof of correctness after each mathematically based transformation, they can theoretically hold defects to zero. Even if the proof process were not perfect, it would reduce defects to levels much lower than today's practice.

Better quality

Quality ultimately rests on the satisfaction the user takes in the software product. This satisfaction depends, in the first instance, on the product meeting the user's needs. Meeting these needs, in turn, is what the requirements process is trying to achieve. Then, as additional needs develop in operation, various enhancement and release procedures carry the process of meeting needs forward. As you improve the process of determining requirements, you also improve product quality.

In the second instance, satisfaction depends on the reliability of the product. Does it operate long enough between failures to satisfy the needs of its operating environment?

When you make these ideas work, your cost of producing software will actually decline. That is where the thought "quality is free" comes from. For most of us, that day is not yet here. In effect, you now have another tradeoff to make—how much can you afford to invest this year to realize "quality is free" in a later year?

Much of the effort devoted to new systems is spent trying to understand the information needs of the enterprise and get them solidified in requirements and specifications. This endeavor is business-oriented, not software-technical. Executives certainly understand the business end of it. Formal specifications, design, coding, and testing get more technical and are left to the technologists. —Vice president of finance for a high-technology company

Reference

[1] F.P. Brooks Jr., *The Mythical Man-Month: Essays on Software Engineering,* Addison-Wesley Publishing Co., Reading, Mass., 1974, 195 pp.

Chapter 27

Coping with Maintenance

Corporations rely on maintenance to keep their daily operations running smoothly, to keep software in a state that allows easy modification and reuse, and to facilitate strategic corporate transition to new software systems. — Robert S. Arnold and Roger J. Martin. [1]

Software organizations spend from 50 to 80 percent of their budgets on what they loosely call "maintenance." One authority even mentions 95 percent. Whatever the exact figures, we spend a lot of money in "them thar maintenance hills" [2, 3].

What is it?

An innocent might say, "The instruction is the unit of software, like a nut or bolt is the unit of hardware. If a nut is defective, replace it with a new nut. If an instruction is defective, replace it with a new instruction."

A sophisticate might say, "Maintenance is not just replacing one instruction; it is everything you do to software after you deliver it and that's a lot."

Indeed, maintenance has come to stand for a series of activities, including

❏ Correcting an error after delivery anywhere in the string of development activities—requirements analysis, specifications, design, or code—without intending to change the intent of the program.

❏ Adding to or modifying the requirements analysis, specifications, design, and code to accommodate changes in the environment in which the program operates. The changes may be either in the human or institutional environment, like new laws, or changes in the hardware environment.

❑   Adding to or modifying the system to enhance its performance, to add functions, to improve the user interface, and so on.

Beyond this stage comes starting over again, replacing the old, now badly deteriorated system with a new one. You are going to redevelop all its functions, usually with new hardware, new tools, new language, and lots of entirely new functions.

Why is it so difficult to "understand"?

In many cases design documentation is not available. In a few cases the developers never reduced requirements, specifications, and design to writing. After some talk among themselves, they simply went directly to source code.

In many more cases they prepared this kind of documentation, but did not keep it up to date. Now only the source code is dependable, and sometimes, if developers have patched object code directly into memory, only the executable code is reliable. In these cases the original, but now dated documentation, may be helpful as a guide to the general direction taken, but it seldom explains why developers followed one path and not others.

Thus, one reason maintainers find it difficult to maintain software is that they find it hard to understand.

It is easy enough to understand a single instruction. There is a paragraph in a computer or language manual somewhere that states precisely what an instruction causes the computer to do. It may be easy for an experienced programmer to grasp what half a dozen instructions are doing. Thousands of instructions are another matter. He must build up levels of understanding from the individual instructions to a grasp of what a module is doing, and eventually to some grasp of what the program as a whole does.

Moreover, there may be difficult-to-understand idioms scattered throughout the code "that correspond to idiosyncrasies in the current software environment," according to researchers at the University of California, Irvine [4]. "The code for even the most straightforward algorithms is almost always disguised by optimizations that depend heavily on both local context and on (inaccessible) global design decisions. Code for complex algorithms is nearly impossible to disentangle."

The original developers understood these complexities. Maintenance is usually the task of new people who did not work on the original development. They lack the understanding that comes with working on a complex project for several years. The new people must depend on the documentation, such as it is. Even if it is clear, there may be thousands of pages of it—too much to absorb.

To work expeditiously, they try to pick out just what they need to know. They pass by hundreds of pages that seem to be irrelevant to their immediate concern. Unfortunately, somewhere in these bypassed pages may be a detail that pertains to the correction they are trying to make.

Why do programmers introduce new defects?

The simple answer is they don't fully understand the code they are trying to correct.

"The more complete and correct the understanding, the more likely that the modifications will be correct," according to researchers at Yale University [5]. "Typically, however, the maintainer is under pressure to carry out the modification as quickly as possible."

Maintainers tend to concentrate their attention on their immediate assignment. Often they form "only a local, partial understanding of the program." With this limited grasp they fail to realize, for example, that the parts of the code they are working on have some, perhaps obscure, interactions with distant other parts. The unfortunate result is the creation of errors in the new or modified code.

How do errors get corrected?

Not surprisingly, in view of the wide spectrum of activities that maintenance covers, planners estimate the size of a maintenance project (and the time and effort) in different ways at different parts of this spectrum.

As we saw in the last chapter, when a program reaches full operational capability, some errors are still there; that is, the Rayleigh curve of defects extends beyond that milestone. The curve lets managers estimate the calendar time and associated effort to reduce the remaining errors, or to increase reliability to 99 percent, or 99.9 percent.

A software organization may extend the date of delivery beyond the fully operational milestone, striving to attain higher reliability before placing the product in the hands of users. In that event the time and effort budgeted for defect removal after delivery will be less. Also, if the software organization uses the quality-improvement measures we outlined in previous chapters, the remaining defects will be fewer; that is, the Rayleigh defect and effort curves will be smaller.

Essentially management has the means to estimate the level of effort needed to correct the errors remaining at the time of delivery. Suppose five defects per thousand source lines of code remain in a 100 KSLOC program, which translates to 500 total defects. That number is large enough to estimate with statistical reliability the time and effort required to find them.

By dint of appropriate tradeoffs, defect-prevention methods, and extending defect-removal efforts beyond full operational capability, a project may deliver its product with 0.5 defects per thousand source lines of code, that is, with only 50 defects remaining. Now we are getting down to where statistical projections of just when users will discover these defects become less reliable. Consequently, estimating the time and effort to find these few remaining defects is less certain.

When is it modification?

Beyond the initial error-correction period is the time when program modification begins. Of course, there can be no guarantee that users have discovered all the errors. As Figure 27-1 shows these modifications then become an increase on top of the tail of the original Rayleigh curve.

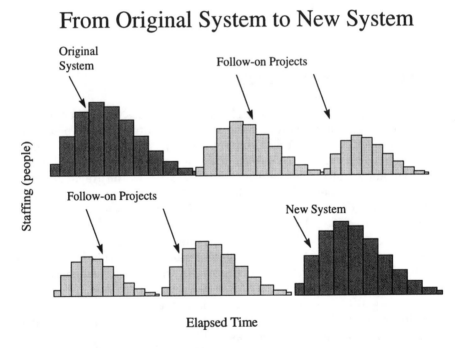

From Original System to New System

Figure 27-1. The Rayleigh curve for the original development terminates in a long tail representing a level of error-correction effort. Periodically during the life of the system maintainers can estimate major enhancements as if they were new projects. Eventually a new product replaces the deteriorated system.

These modifications are new work, something more than just fixes to the original code. They are enhancements to the original requirements or changes to adapt the code to new environmental circumstances. From an estimating standpoint, they fall into two categories.

1. If the amount of work is sufficient to be a project in itself, planners can estimate the size and compute the management numbers. Work of this sort is a project if it is cohesive or connected, if it involves the normal project stages from requirements analysis to testing, and if it is expected to take six or more months of development time, more than 20 person-months, and three people or more. If the work is not "connected," but consists of little, isolated tasks, it is not suitable for estimating in this way.

2. When there are many of these isolated tasks, managers can estimate a level of effort on the basis of ongoing experience. They assign enough people to keep up with the rate of incoming enhancements and changes. If the backlog becomes too great, they assign more people.

Maintenance tasks in these two categories usually involve another complication: whether the code is new or modified. An addition to the existing system may well be in the form of new modules. Planners can estimate the size of the new code and calculate time and effort in the same way as they would for a new project.

Modifications to existing code consist of changing code or perhaps adding a few lines in an existing module. They are generally less time-consuming than developing an entirely new module of the same size. Some organizations have developed methods of estimating the effort involved by saying, in effect, modifying this module will take 30 percent of the effort of writing an entirely new module. You can develop ratios of this sort from your experience.

When is it time to replace the system?

Eventually your environment will have changed substantially. You will be on your second or third hardware generation since installing the original software. You will have patched that software thousands of times. It will have deteriorated into an impenetrable mess that causes good programmers to go elsewhere.

It will be time to start from scratch—reanalyze requirements, rewrite specifications, develop an entirely new system. From an estimating standpoint, it is a new system and requires the estimating methods applicable to such systems.

Since [maintainers] cannot possibly have as high a level of product knowledge as the designers do, the system should be kept simple and clearly documented: the more unstructured the system, the more deterioration of structure will take place during maintenance. —Les Belady [6]

References

[1] R.S. Arnold and R.J. Martin, "Software Maintenance," *IEEE Software*, May 1986, pp. 4–5.

[2] D.H. Longstreet, *Software Maintenance and Computers*, IEEE Computer Society Press, Los Alamitos, Calif., 1990, 294 pp.

[3] B.E. Swanson and C.M. Beath, *Maintaining Information Systems in Organizations*, John Wiley and Sons, New York, N.Y., 1989.

[4] G. Arango et al., "TMM: Software Maintenance by Transformation," *IEEE Software*, May 1986, pp. 27–39.

[5] S. Letovsky and E, Soloway, "Delocalized Plans and Program Comprehension," *IEEE Software*, May 1986, pp. 41–49.

[6] M.M. Lehman and L.A. Belady, *Program Evolution: Processes of Software Change*, Academic Press, New York, N.Y., 1985, 538 pp.

Chapter 28

Better, Faster, and Cheaper

*"Do it better, faster, and cheaper. That's the implicit signal we're get-
ting from corporate,"* the division general manager told us. *"The corporation is under increasing competitive pressure. Strong competitors
are springing up all over."*

The division is a unit of a major manufacturer. The unit manufactures
telecommunications equipment and develops the software embedded in
the equipment. The division had been trying several software cost-
estimating packages and was beginning to have confidence in the results.

"Corporate has seized on the idea of market-driven quality," the general manager continued. *"It's a great idea. Of course, we should pay
attention to our customers. Of course, we should give them quality
that meets their needs."*

*"We are quite taken with the ideas of total quality management our-
selves,"* we declared.

"Now we have this explicit directive from corporate," he went on. *"Last
year the corporate standard was 1.5 defects per thousand source lines
of code, remaining in the product after delivery."*

"That is a very good level," we said.

*"Yes, but now corporate says, 'You will have a tenfold increase in
quality in two years.' The new goal is 0.15 defect per thousand source
lines of code."*

"That is very ambitious," we agreed. *"The few organizations we've
heard of that have reached that level seem to have devoted a great
deal of time and effort to getting there."*

*"You recommend reducing errors by allowing more development
time,"* the general manager went on.

"Yes, allowing a reasonable development period does reduce defects," we assented.

"Ah ha! The product of that tradeoff is better and cheaper, but not faster."

"That's right," we agreed, *"but management at the working level has to make tradeoffs. Maybe at the corporate level they don't have to get their feet muddy."*

"I think I'll put my rubbers on," he smiled. *"We are going to take advantage of this tradeoff. We are investigating schedules on forthcoming products to see if we can get software started a little sooner and allow a little more development time."*

"Good idea," we murmured.

"At best, however, this tradeoff will not enable us to meet the new corporate quality goal. Even if we plan schedules 130 percent longer than minimum, we fall far short."

"Yes, we went through it with your software people," we said. *"The new standard will be very difficult to reach on the process productivity index on which you are now operating—around 10. The standard probably came from the MIS or data-processing world, in which the average productivity index is around 16. In the engineering applications world (command and control, systems software, telecommunications, process control, and scientific software), the work is more difficult. The process productivity indexes tend to center around 9 to 12."*

"My people say it will take an index of at least 15 to drive reliability that high, without trading time for fewer errors."

"Yes, time stretchout and staff reduction can get you to this reliability goal, but only if you are at the right productivity index," we explained. *"If you can't reach a high quality goal with this approach, it is because your productivity index is too low for the standard. Or we might say the standard is not appropriate for your current level of process productivity. There is a solution: attack the process with an improvement program that moves you up the productivity scale. Unfortunately, this approach is rarely a short-term proposition."*

"Is there anything we can do in the short term?"

"Yes, there is, but it isn't glamorous. It is just plain old 'work out the errors' until you get to the standard required," we answered. *"You have to work beyond full operational capability into the maintenance phase before delivering the product. It takes more time and effort."*

"Quite a bit more if I remember correctly." he said. "Your rule of thumb is 50 percent extension of development time to get to 99.9 percent reliability."

"Right, it's better, but not cheaper and not faster," we admitted.

"Maybe I'll get some rubber boots," he said ruefully.

"It takes time to find defects and then it takes more time to correct them. It takes thought—and time—to avoid introducing more errors in the process," we noted. "Extending delivery time provides the time for this. The common denominator here is, if numerous errors creep into the software, it takes time and effort to get them out."

"I think I see what you are driving at: keep the errors out of the software in the first place," the general manager said. "Then we can have better quality, faster and cheaper."

"Yes, that is the only way to get all three," we agreed.

"Someone showed me a table the other day of 'cleanroom' projects," he observed. *"Two of them, in the 50,000 SLOC range, had no failures in operation. The rest were down in fractions of a defect per thousand source lines of code."*

"Right, some people are making it work."

"One point, I still don't understand," he went on. "My software director showed me a diagram [Included here as Figure 28-1] in which mean time to defect increases rapidly as the productivity index improves. Do you mean to say that bettering productivity also betters quality?"

"Yes, the two go together. If you take steps to prevent defects, productivity—that is, process productivity—improves. If you take steps to improve process productivity, defects decrease and MTTD increases."

"I would think off-hand that all the time spent preventing defects would reduce productivity," he objected.

---

* "Cleanroom software engineering" is the name Harlan D. Mills gave to the methods he developed at IBM to produce defect-free software. Briefly, Cleanroom engineering bends every effort to get a correct, formal specification. It then transforms the specification into code through a series of mathematically based operations that greatly reduce the incidence of errors.

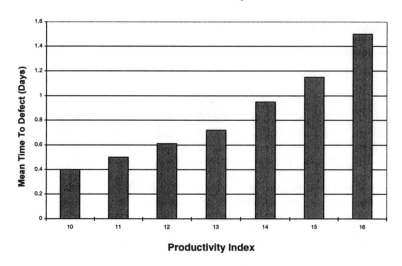

MTTD vs. Productivity Index

Figure 28-1. How mean time to defect relates to the productivity index. MTTD (at the end of development) increases rapidly with process productivity (other factors remaining constant). This chart is based on the main build of a 100,000 SLOC telecommunications application at a manpower buildup index of 3.

"Suppose it does take more time and effort to apply some kind of disciplined approach to preventing errors," we answered. "Later you save even more effort in finding and fixing errors. Remember, it takes perhaps 100 times as much effort to find and fix an error late as early."

"That makes sense."

"Besides, it probably doesn't take more time and effort to be disciplined," we added. "The heuristic approach leads to false starts and wrong directions. A disciplined approach might be a little slower to start but it moves more surefootedly."

"That seems likely enough."

"Remember, we are referring to process productivity, not conventional SLOC/person-month productivity," we continued. "Process productivity encompasses management practices, organizational capability, people skills and experience, the state of equipment technology, and software tools and methods. Most of these factors help reduce errors as well as increase productivity in the conventional sense."

"Unfortunately, it looks like improving process productivity doesn't work fast enough for us to meet the corporate quality goal," the general manager noted.

"Afraid not. Our database shows that for systems of your application type the average rate of improvement in the productivity index is one point every three years. So it might take 15 years to reach the corporate quality goal by that route, not the two years corporate wants."

"I had hoped we could do better than average," he replied, "but two years sounds like blue sky."

"Everybody else—even corporate—is up against the same facts," we pointed out. "This chart [Included here as Figure 28-2] shows how we expect process productivity to improve in the future. It gives you an idea of what a long, slogging trail it is."

"Let me think through all this," the general manager said after a few moments of reflection. "Efforts to improve quality quickly, as in the two-year goal set by corporate, involve a time tradeoff that violates corporate's implicit 'faster' goal. We can improve quality without sacrificing the 'faster' goal only by improving process productivity. Bettering process productivity also reduces defects. We can make some progress along this improvement trail in two years, but substantial progress stretches out over many years."

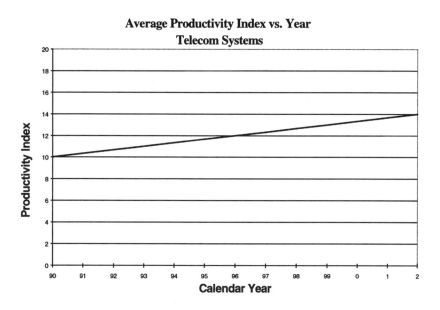

Figure 28-2. Process productivity of telecommunications systems may improve over the 1990's decade as shown if the past rate of average improvement—one index point every three years—continues. This chart represents a statistical average of many companies slogging through their work, often without a real process-improvement program. You can do better if you have such a plan. See Part V.

"You've put it well," we said. "Corporate must set goals that reflect the kind of work each division is doing. Most important, corporate must set in motion a long-term process-improvement and defect-reduction program. Only in that way can it get 'better, faster, and cheaper.'"

Part V

Process Improvement Phase

> *"Process improvement is based on fundamental concepts, first developed by Walter Shewhart and later by J.M. Juran and W. Edwards Deming, of 'process control'. The nature of a process is dynamic, that events happen in a time sequence, and 'flow' through various stages."*
> —Tom Gilb and Dorothy Graham [1]

Only by improving the process can you improve all the management numbers at the same time—schedule, staffing, effort, cost, reliability. Only by investing in equipment, tools, and the education and training of staff and management can you improve the software development process. Only by having a measure of the reality of the ensuing improvement can you be comfortable in continuing to invest in it.

Only by successfully investing in process improvement can you keep up with other software organizations that are improving their processes year after year. Process improvement can be planned and accomplished, as organizations reporting to our database demonstrate.

In fact, process improvement is itself a process. As a process, if it can be measured, it can be subjected to dynamic control like any other process.

That is the essential purpose of this Part: to show that process improvement can be measured and controlled.

> *"The process we are interested in getting under control is our own software engineering process."* —Tom Gilb and Dorothy Graham [1]

Reference

[1] T. Gilb and D. Graham, *Software Inspections*, Addison-Wesley Publishing Co., Reading, Mass. 1993, 471 pp.

Chapter 29

A CEO Thinks About Process Improvement

"Your budget request for the coming year contains $200,000 for software process improvement," the chief executive officer observed, looking at the page of figures in front of him.

The director of software development nodded. "It's about the same as last year."

"Well, capital funds are scarcer than usual this year," the CEO said. "Why should we spend that much on the software process? You've been investing money in workstations, software tools, and the like for years. Yet most of your jobs are still over schedule and over budget and riddled with errors." The CEO paused. He felt it was a good thrust.

"I could tell you that the business runs on software, that we must have it, problems or not," the director responded. "But you know that. I could tell you that we are producing more software in less time, but the work keeps getting more complex."

"Yes, you told me that last year," the CEO recalled. "What have you got for me this year?"

"I have the software process productivity index," the director came back. "It is a numerical measure of the effectiveness with which we produce software. It is not based on my judgment or on some kind of an evaluation by consultants. It is calibrated from our records of project size, schedule duration, and people effort. Hence it is objective."

*"Now that I am interested in." The CEO leaned forward. "One thing I learned in college was Lord Kelvin's famous statement about measurement."** *

"I looked at where we have been," the director said. "I put the project management numbers for the last five years through the calibration procedure. Here are the average productivity indexes of the projects completed each year:"

1989	10.7
1990	11.0
1991	11.3
1992	11.7

"That rate of advance isn't earth-shattering," the CEO said, frowning. "Still, it shows steady progress. Moreover, at least it gives us some numbers to work with."

"That's true, they are numbers and they are increasing," the director agreed, "and you put your finger on the problem. The numbers themselves are mediocre. In business systems the average productivity index is 16. So we are well below average. In fact, the standard deviation of the average productivity index is four points. So we are at the top of the bottom 16 percent." [When the two executives were talking, the productivity index was 16.0 ±4. The latest figure is 16.9 ±4.9.]

"That describes a guy who might get a cost-of-living increase, but a bonus—ho! ho!"

"Very funny, but the index people warn us not to mix production measurements with personnel evaluations," the director said. "People tend to distort metrics in that case."

"Just kidding," the CEO admitted. "In all seriousness, though, being way below average is not good enough in this dog-eat-dog world."

"It kept me awake last night—I got these figures only late yesterday."

"We've been gaining about one third of an index point per year, or a point every three years," the CEO mused.

---

* *"When you can measure what you are speaking about, and express it in numbers, you know something about it; but when you cannot measure it, when you cannot express it in numbers, your knowledge is of a meager and unsatisfactory kind."* —Lord Kelvin, Popular Lectures and Addresses, 1889.

"That doesn't sound like much," the director said, "but it amounts to an improvement in process productivity of more than eight percent per year, compounded annually—and that's not too shabby."

"I wish the rest of the company were doing that well," the CEO added. "Still, at a point every three years, it would take us about 12 years to get up to average."

"That average is a moving target, too," the director admitted. "It's going up a little faster than we are."

"I see, things get a little complicated." The CEO let his thoughts run on. "At least with this productivity index we have an idea of where our software development stands. We know we must do better than we have been. And we have a metric that will tell us whether we are."

"The metric will clue us on whether the money we invest in process improvement is working," the director added.

"Obviously. What we have been doing has been good for only a third of a point a year," the CEO said. "Doing better probably will take more. Leave your papers with me so that I can come up to speed. We'll talk about it Thursday."

Chapter 30

Process Productivity Marches On

"The relative strengths of the leading nations in world affairs never remain constant, principally because of the uneven rate of growth among different societies and of the technological and organizational breakthroughs which bring a greater advantage to one society than to another." —Paul Kennedy [1]

We might echo that: The relative strengths of the leading organizations in software affairs never remain constant, principally because of the uneven rate of growth among different companies and of the technological and organizational breakthroughs, which bring a greater advantage to one company than to another."

Paul Kennedy was examining 500 years of history. In software the pace of change is more like 500 weeks. In this chapter we look at where process productivity is now, the progress it has made in the last 500 weeks, and where it might go in the next 500 weeks. We shall find that process productivity has been a moving target and it promises to continue moving. Organizations must run hard just to keep in the same place.

The productivity record

The software organizations reporting data to us did improve their process productivity substantially during the 1980s. Remember, "process productivity" is a broad term—much more inclusive than conventional SLOC/person-hour productivity. It includes not only all the factors affecting an organization's ability to produce software, but also the complexity of the work itself.

Differing degrees of complexity are the main reason systems in different application areas fall at different levels on the productivity-index scale, as listed in Table 6-1. Similarly, different application areas have reported different rates of improvement in process productivity, as shown in Table 30-1.*

The "Best sustained record" in the first row of the table refers to an organization that has been improving its process productivity consistently for the 15 years we have been following it. This record shows what is possible; it represents a practical goal.

There is one more disclaimer. We derived the figures in the table entirely from our database. This database, large as it is, is not necessarily representative of the entire universe of software organizations. We believe that organizations reporting data to us are more efficient than organizations in general. Organizations that are not paying consistent attention to process improvement may be remaining static or improving by only a few percent per year.

Table 30-1. Process productivity increased during the 1980s as measured by our productivity index. In percentage terms (last column) the gain (in the process productivity parameter) has been impressive.

Application	Rate of Improvement in Process Productivity		
	One PI Every	**PI/Year**	**Percent/Year**
Business Systems			
Best Sustained Record	1.5 Years	0.67	16
Average	2.5 Years	0.40	10
Engineering Systems*	3.0 Years	0.33	8
			8
Real-Time Systems	4.0 Years	0.25	6

*Includes command and control systems, systems software, telecommunications systems, process control, and scientific software.

---

* The process-productivity index numbers are a linear scale that represents the process-productivity parameter, as Figure 6-1 shows. The process-productivity parameter increases at an exponential rate. Hence, to show the rate of improvement only in terms of the index numbers does not express the real rate of increase. Therefore, the column headed "Rate of Improvement in Percent per Year" is based on the percentage increase in the process-productivity parameter. This column is a realistic expression of the overall gain in effectiveness of the organizations reporting to our database.

Potential gains

Because many software organizations are not up to the level at which they are collecting good data, they can enjoy great improvements—if they care to exploit the possibilities. Even the more efficient organizations represented in the database can improve very substantially, as Table 30-2 shows. The first column contains the current average productivity index; the second column, the standard deviation in index numbers. Now some of these more efficient organizations, of course, are still below our average. What would it mean for one below average (say by one standard deviation) to improve to above average (one standard deviation above average)?

The third column shows this improvement factor. This factor is the ratio of the process-productivity parameter value one standard deviation above the mean to the value one standard deviation below. For real-time systems, for example, the parameter values (which we took from Table 6-1) are 8,620/1,559. This ratio represents a potential improvement factor of 5.5.

In other words, an organization advancing from the 16th percentile to the 84th improves its process productivity by this factor. This degree of improvement seems entirely reasonable—obviously well within the state of the art, since many organizations are already operating at the higher level. They know how to do it. Your organization can learn how, if it isn't already there.

Moreover, some organizations are more than one standard deviation below the mean. They stand to gain even more than the improvement factors shown in the table if they can pull themselves up above +1 standard deviation.

Table 30-2. Process productivity improvement potential. An organization now one standard deviation below the mean can increase its process productivity by a factor of five to 11 by raising its productivity to one standard deviation above the mean.

Application	Average Productivity Index	Standard Deviation	Improvement Factor
Business Systems	16.9	4.9	10.6
Engineering Systems	9.0	3.7	5.9
Real-Time Systems	7.5	3.6	5.5

A few organizations are reporting productivity indexes on some systems in the upper twenties and low thirties. These indexes constitute an existence proof that considerable room for improvement lies ahead of nearly all software organizations.

The decade of the 1990s

If the organizations in our database continue to improve during the 1990s at the rate of the previous decade, the gains in their productivity indexes will be as shown in Figures 30-1 through 30-4.

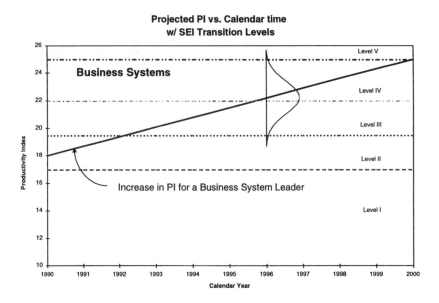

Figure 30-1. The business software organization with the best sustained record may improve from a productivity index 18 to 25 over the decade of the 1990s.

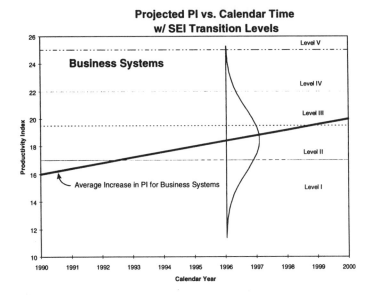

Figure 30-2. The average business organization, improving at a slower rate than the leader, will fall farther behind the leader. The normal curve, superimposed on the trend line, shows the 1996 distribution of productivity-index values above and below the trend line. On an interactive display, a button on the normal curve enables a viewer to move it up and down the trend line.

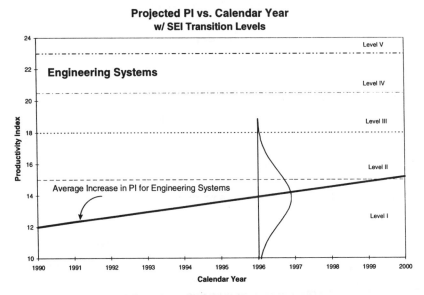

Figure 30-3. Engineering systems are more complex than business systems, so their average productivity index will probably lag behind that of business systems. These diagrams reflect our database, not the software community in general.

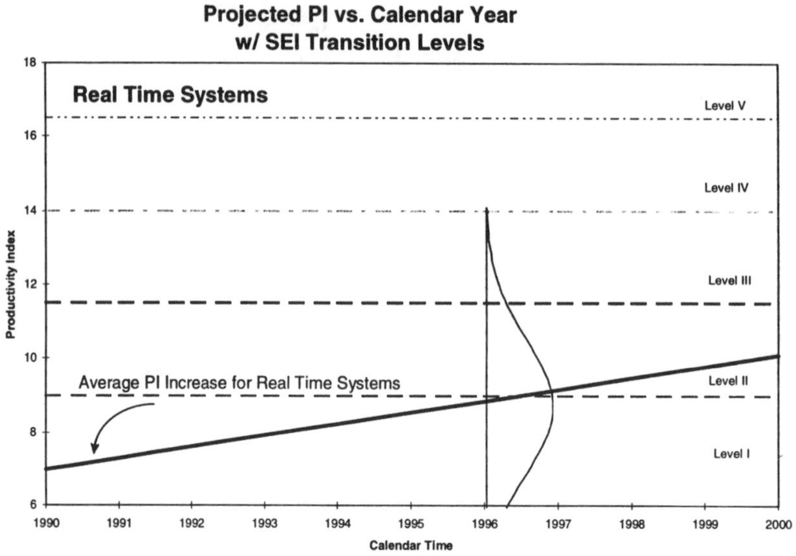

**Projected PI vs. Calendar Year
w/ SEI Transition Levels**

Real Time Systems

Level V

Level IV

Level III

Productivity Index

Average PI Increase for Real Time Systems

Level II

Level I

1990 1991 1992 1993 1994 1995 1996 1997 1998 1999 2000

Calendar Time

Figure 30-4. Real-time systems are the most difficult application category of all, as shown by this trend line.

On these diagrams we superimposed the software-organization maturity levels devised by Watts S. Humphrey of the Software Engineering Institute of Carnegie Mellon University [2]. The levels are defined below. After four years of experience with this process-maturity framework, the SEI evolved a modified version called the Capability Maturity Model, Version 1.1 [3].

1. *Initial.* "The organization typically operates without formalized procedures, cost estimates, and project plans. Tools are neither well integrated with the process nor uniformly applied. Change control is lax, and there is little senior management exposure or understanding of the problems and issues. Since many problems are deferred or even forgotten, software installation and maintenance often present serious problems."

2. *Repeatable.* "The organization has achieved a stable process with a repeatable level of statistical control by initiating rigorous project management of commitments, costs, schedules, and changes." In other words, it can repeat its existing way of working on a new project of the same type. However, organizations at this level "face major risks when they are presented with new challenges."

3. *Defined.* The key actions required to advance to this level "are to establish a process group, establish a development process architecture, and introduce a family of software engineering methods and technologies." However, this level "is still only qualitative: there is little data to indicate how much is accomplished or how effective the process is."

4. *Managed.* "The organization has initiated comprehensive process measurements and analysis." It has the measurements of its own behavior needed to manage the process. Characteristics of this level are "a minimum basic set of process measurements to identify the quality and cost parameters of each process step, ...a process database, ...resources ...to advise project members on its use, ...[and] an independent quality assurance group."

5. *Optimizing.* "The organization now has a foundation for continuing improvement and optimization of the process." There is a paradigm shift. "Up to this point software development managers have largely focused on their products and will typically gather and analyze only data that directly relates to product improvement. At the Optimizing level "the data is available to tune the process itself."

As of October 1990, of the 113 project assessments conducted or observed by the SEI, 85 percent fell into level 1, 14 percent were at level 2, 1 percent at level 3, and none in levels 4 or 5," John P. Murray reported [4].

The point the diagrams make is that on the average it will take a long time, and no doubt a great deal of investment and hard work, to gain a single SEI level. "Achieving higher levels of software-process maturity is incremental and requires a long-term commitment to continuous process improvement," SEI authors stated. "Software organizations may take 10 years or more to build the foundation for, and a culture oriented toward, continuous process improvement" [3].

Two years later an SEI report stated that, as of December 1992, none of 150 assessments and seven reassessments of government and industry sites had attained level 5 (optimized) or level 4 (managed). Seven percent were at level 3 (defined), 19 percent at level 2 (repeatable), and 74 percent at level 1 (initial). The sites assessed were mostly defense-related: defense contractors, 41 percent, and military, 16 percent. The rest were commercial, 34 percent; government, 7 percent; and other, 1 percent [5].

In September 1994 an SEI group reported the results of 284 assessments: 0.5 percent at level 5 (optimized), 0 percent at level 4 (managed); 8 percent at level 3 (defined), 16 percent at level 2 (repeatable), and 75 percent at level 1 (initial) [6]. "This distribution has not changed substantially since the last figures were published," the group said. "This is not surprising, since a very high percentage of these assessments (261 of 284) are first assessments." However, "of the 23 organizations that have had two assessments, 83 percent have moved up a maturity level, while the remaining 17 percent stayed the same." There was an average of 27 months from the first to the second assessment. Evidently the interest generated by an assessment led in most cases to a successful effort to move up the Capability Maturity scale.

In April 1995 the results of 435 assessments performed through December 1994 again showed little change:

Level 5	0.3 percent
Level 4	0.6 percent
Level 3	10 percent
Level 2	16 percent
Level 1	73 percent

As has been the case all along, the great majority of software organizations continue to rest on the lowest level. A few, however, are beginning to penetrate the upper levels, demonstrating, by example, that this improvement is possible.

What effect does the prospective improvement in process productivity have on the management numbers? Let us take a typical business system of 70,000 SLOC and a middle-of-the-road manpower buildup index of 2.5. If the average rate of improvement during the 1980s continues for another decade, the productivity index will advance from an average of 16 in 1990 to 20 in 2000. Effort, schedule, and defects remaining per thousand source lines of code will decline, as Figure 30-5 shows. The vertical scale is logarithmic, so the rate of decline in linear terms is much greater than it appears on this figure. The mean time to defect increases.

References

[1] P. Kennedy, *The Rise and Fall of the Great Powers,* Random House, New York, N.Y., 1987, 677 pp.

[2] W.S. Humphrey, *Managing the Software Process*, Addison-Wesley Publishing Co., Reading, Mass., 1989, 494 pp.

[3] Paulk, et al., "Capability Maturity Model, Version 1.1," *IEEE Software*, July 1993, pp. 18–27.

[4] J.P. Murray, "Software Development Process Maturity," *System Development*, May 1991, pp. 8–9.

[5] J.H. Baumert and S.M. Howard, "SEI Hosts An Open, Useful Symposium," *IEEE* Software, Nov. 1993, pp. 101–102.

[6] J. Herbsleb et al., "Software Process Improvement: State of the Payoff," *American Programmer*, Sept. 1994, pp. 2–12.

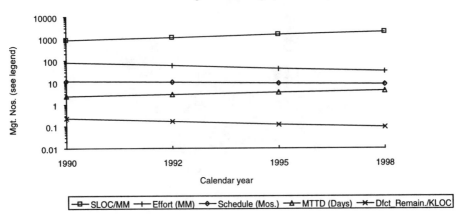

Figure 30-5. If process productivity continues to improve at the average rate achieved in the 1980s, the management numbers should improve during the 1990s

"There's one achievment I'm proud of, Bernard. We never abased out creativity with 'comprehensive process measurements.' "

Chapter 31

Process Improvement
Yields Results

"When a process is under statistical control, repeating the work in roughly the same way will produce roughly the same result. To obtain consistently better results, it is thus necessary to improve the process."
—Watts S. Humphrey [1]

One productivity index point every two to three years, as reported in the previous chapter, may sound discouragingly slow. Keep in mind, however, that software development is a very difficult activity. It interfaces to the real world on one side—which is as complicated as anything short of the galaxy itself—and to computer, peripheral, and communications hardware on the other side—and they too are complicated in a different way.

Software development uses human beings to create software, and, well, they have 100 billion neurons each. Anyone who has tried to understand another human being knows just how complicated 100 billion neurons can get.

Then we put all this complexity together in organizations with several levels of management. That compounds the complexity. On top of that the organization and many of its denizens tend to resist change. But you know all this. You push and push, and the organization moves an inch.

Our point is this: It is worth pushing process improvement. The gains are larger and come sooner than you may have expected. Moreover, if you push skillfully, you can gain process productivity more rapidly than average.

What you gain

When you improve process productivity, you shorten your schedule; reduce effort, cost, staff, and defects; and increase mean time to defect. To demonstrate these effects, we chose a typical 100,000 SLOC business system. We held the manpower buildup index constant near the middle of the scale at 3.0. We then let the productivity index increase from 10.0 to 22.0. In this way we could observe what happens to the management numbers as this organization becomes more effective. This index range is realistic. About 80 to 85 percent of the systems in our database fall within it.

❑ *Schedule.* First we looked at schedule. Figure 31-1 shows that the length of time to develop this system declines markedly as the organization improves its process productivity. This gain accomplishes the often-stated goal of "Reducing Cycle Time."

❑ *Effort and cost.* Figure 31-2 shows that the decline in effort (person-months) and cost (millions of dollars) is much more rapid than that for schedule time. Both figures show that you do not have to go from low to high productivity immediately. Realistically that is not possible. They do show that you reap very significant rewards at each stage from 10 to 22.

❑ *Staffing, defects, and MTTD.* As you improve your process, the same amount of function can be generated by smaller teams of people, as Figure 31-3 illustrates. Staffing declines in the same way as effort and cost. As we described in earlier chapters, fewer people, along with improved productivity, produce fewer defects. Fewer defects in turn cause mean time to defect to increase, which the figure also shows.

It may help to see the relationship between effort and reliability. This we do explicitly in Figure 31-4. Effort, defects remaining at delivery, and defects remaining at delivery per thousand source lines of code all decline at the same rate as process productivity improves. In effect, this figure says effort is proportional both to the number of defects created (defects per thousand source lines of code) and to the defects remaining. Note the logarithmic scale on the vertical axis of the figure. Use of this scale turns the bars of Figures 31-1 through 31-3 into straight lines, letting us compress a great range of values into a single diagram.

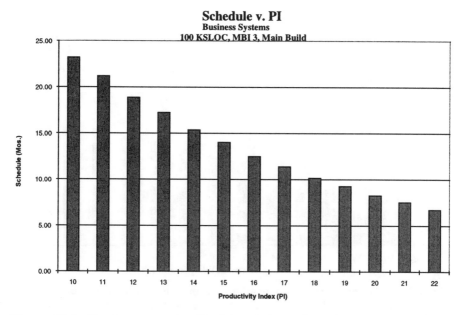

Figure 31-1. Greater process productivity shortens the schedule. (The chart is based on a 100,000 SLOC business system with a manpower buildup index of 3.)

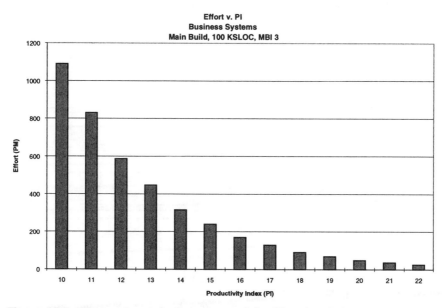

Figure 31-2. Effort and cost decline exponentially. (The chart is based on a 100,000 SLOC business system with a manpower buildup index of 3.)

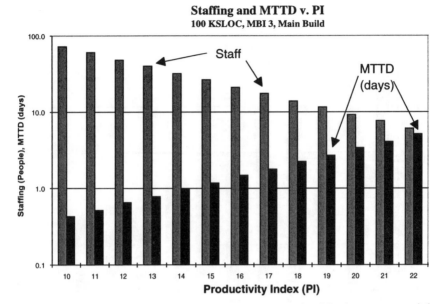

Figure 31-3. Fewer people are needed as process productivity improves, and they make fewer errors, so the MTTD gets longer. (The chart is based on a 100,000 SLOC business system with a manpower buildup index of 3.)

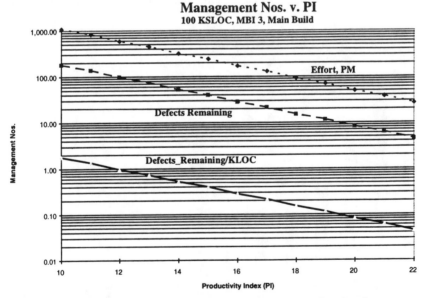

Figure 31-4. Better process productivity not only reduces effort but improves reliability. Note the logarithmic scale on the vertical axis. Use of this scale turns the curves of the preceding figures into straight lines. Its use lets you compress a great range of values to a single diagram. (The chart is based on a 100,000 SLOC business system with a manpower buildup index of 3.)

In Figure 31-5 we have another view of what "Reducing Cycle Time" means. This figure is a representative staffing profile of the same project—a 68,380 SLOC, business system—carried out at low, medium, and high process productivity. Its size is the industry average for business systems in the early 1990s. The diagram includes the high-level functional design phase. The manpower buildup index is 2.5 for each level of process productivity. The three productivity indexes are 16 (mean of business systems in the early 1990s), 12 (one standard deviation below the mean), and 20 (one standard deviation above the mean). The area under the curves is the effort or cost.

As the figure shows, at high productivity you can get in another project of the same size during the time the low-productivity organization would still be struggling with its first project. At high productivity you free up enough people to staff about 10 more similar projects. MTTD is more than five times better for the high-productivity producer than for the low-productivity producer. At high productivity users get the economic benefit of new software months sooner. At high productivity software people can make a real dent in the backlog of work waiting to be done.

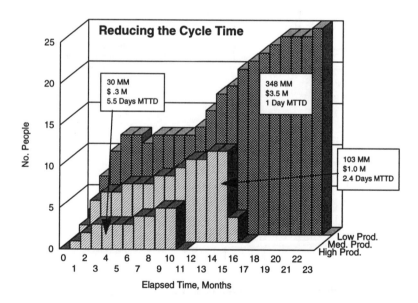

Figure 31-5. The tiny area on left is the high-productivity scenario. It represents your goal, perhaps some years out, if you are now average. Still if you are now below average (area to the right), the average productivity profile looks pretty good. (The chart is based on a 68,380 SLOC project carried out at a manpower buildup index of 2.5 at productivity indexes of 12, 16, and 20.)y

Figure 31-6 shows, at the top, a defect rate curve (in bar form) before process improvement. It shows, at the bottom, to the same scale, the defect rate curve for a similar project after process improvement. (The diagram represents any software organization in any application area.) The area under the lower curve, proportional to total defects, is smaller than the area under the upper curve. The number of defects remaining at delivery has also declined. Because the projects are the same size, total defects/KLOC and defects remaining/KLOC have declined. Finally, development time up to delivery is much shorter. Time to market is better.

While improving process productivity may seem to go slowly in one sense—it takes some time to improve from one productivity index to the next—the benefits pile up rapidly along the way.

Reducing cycle time means shorter schedule, fewer people, less cost, fewer defects, and higher reliability. It is the embodiment of total quality management.

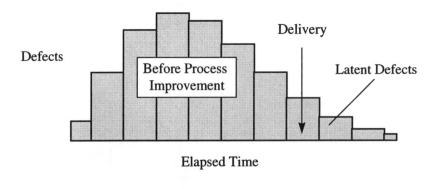

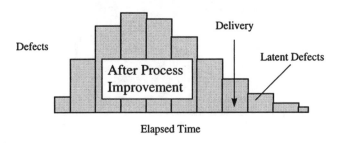

Figure 31-6. Defect-rate curves before and after process improvement demonstrate substantial improvement in total defects, defects remaining at delivery, and schedule.

Raytheon increases productivity

Raytheon's Software Engineering Initiative, begun in 1988, reported improvements of this magnitude in its Software Systems Laboratory with 400 software engineers 4.5 years after Raytheon began the initiative:

- ❑ $7.70 return per dollar invested
- ❑ Factor of 2.3 increase in productivity (measured in SLOC/person-month)*
- ❑ Rework costs reduced from 41 to 11 percent of project costs
- ❑ Two $17 million projects finished ahead of schedule and four to six percent under budget [2]

The Software Engineering Initiative, headed by Raymond Dion, achieved these results over 4.5 years. During this period the laboratory's SEI Capability Maturity Model level increased from an estimated 1 (initial) to an estimated 3 (defined).

"We have recently been awarded at least two contracts in which our software-process capability was said to have made the difference between winning and losing," said Dion.

The bleak side

That was the carrot; now the stick. The cost of runaway or defective systems often gets personalized in the dismissal or demotion of the responsible executive. On the basis of a survey of 600 chief information officers, the consulting and accounting firm, Deloitte & Touche, reported that 25 percent of the CIO's predecessors had been dismissed, 7 percent had been demoted, and 10 percent had retired.

Slightly more than half the 200 chief executive officers and chief financial officers surveyed by *Computerworld* and Andersen Consulting doubted that their companies were getting the full benefit of their computer investments [2].

> *"All the king's horses and all the king's men could not put Humpty Dumpty together again."* But the software process is not an egg. Raytheon—and others—have put it together.

---

* In Chapter 4 we labeled SLOC/person-month a "poor metric." The reason is that it varies with system size and application type. In this case, however, the 18 projects compared are all the same type, real-time embedded, and all rather large, 70,000 to 300,000 SLOC. So, although the factor of 2.3 may be somewhat uncertain because the project size varies, it is surely indicative of a substantial increase in productivity, however it was measured.

References

[1] W.S. Humphrey, *Managing the Software Process*, Addison-Wesley Publishing Co. Reading, Mass., 1989, 494 pp.

[2] R. Dion, "Process Improvement and the Corporate Balance Sheet," *IEEE Software*, July 1993, pp. 28–35.

Chapter 32

Plan to Improve the Process

"I just wanted to ask you which way I ought to go." said Alice. "Well, that depends," replied the Cheshire cat," on where you want to get to."
—Lewis Carroll, *Alice in Wonderland*

"All you guys want to do is spend money on bright ideas," the vice president of operations told us. "I used to have bright ideas myself when I was younger, but now I just try to keep things running somehow."

"We're running a business ourselves," we replied. "We're old enough to know better, too."

"I can see that," he laughed. "All the same, at every management committee meeting our chief financial officer goes over the figures. There is never enough to cover all the things we ought to be doing. In fact, several of our software projects are overrunning their budgets. The whole department is in the red."

"No one ever said running a business was all income," we purred sympathetically.

"Your charts of what people are accomplishing are very seductive," he admitted. "I guess the real question is where do I begin?"

Do what you know you should

In the first place, you start where you are. Because accomplishing a miracle seems impossible, lots of us don't do the little things we could do.

You might take a look at your hiring, training, review, and compensation practices for software people. Everyone agrees that good people, adequately trained, satisfied with their pay, happy in their work, do a better job.

You might take a look their physical working conditions. Requirements analysis, specification writing, detailed design, coding, test planning, testing, debugging—you know this work is difficult. And it's no secret that work calling for hours of concentration can be done more efficiently under conditions of peace and quiet. Sure private offices for everyone would cost a lot of money (which you don't have—we hear you). But you could think through what you can do without money or with only a little money.

By moving bookcases, file cabinets, or blackboards around, you may be able to create a modicum of privacy. Of course, to take action as drastic as this, you may have to outwit Tom DeMarco and Timothy Lister's "Furniture Police." They are the people, not themselves intellectual workers, but who make up the rules governing the working environment [1].

By moving meetings out of the bullpen, you may make it easier for other occupants to get on with their work. By putting the desks of testers, field people, and others who spend much of their time away from their desks in the offices of analysts, designers, and coders, who spend much time at their desks, you may provide more peace and quiet for the desk-bound.

Constantly ringing telephones are not only an annoyance, they interrupt work. Managers have secretaries to protect them from unnecessary interruptions. Ouch! We know, secretaries equal money, too. Can your telephone system intercept calls to professionals who are not even in the plant at the time, saving people at nearby desks from interruption? The administrative services manager could give some thought to this problem.

Electronic mail avoids the ringing telephone irritant, but it introduces a new time waster—being copied with information you don't need to know. Another problem for the administrative services manager to think about.

You might take a look at the wherewithal that people need to do software work: desks, tables, chairs, file cabinets, bookcases, books and manuals, copiers, workstations, printers, electrical outlets, proper lighting, and so on. Do they have them? Sure, they cost money, but not much compared to the cost of professional time. Anyway, you can make incremental improvements as funds permit.

A manager doesn't have to know much about software technology to initiate improvements of this kind that will enhance process productivity appreciably. A manager just has to have the authority to rearrange things a bit or to spend a little money. Managers have that authority. Software people don't. However, they may know the software process in greater depth than managers whose experience is in some other discipline.

Do what software people know

One of the things software people know is that there is a long string of technologies that can increase process productivity.

"We can't do a long string of things," a manager might say. "Remember money. Also, a lot of bright ideas don't work. Some of them may work for other people, but not in our particular circumstances."

These problems are not unique to software. What do people do in other technical areas? What did the managers do about acquiring the "superfragilistic bombardment mechanism"?

They probably did something along these lines. They put together some kind of a task force that brought together the required knowledge. In the software case, the group would include software technologists, representatives of the software people affected by the proposed technology, and perhaps some users.

The managers would then ask the task force to list some promising improvements. They would want the task force to pick items that look like they would work. The software technologists are the ones who ought to know. The managers would have them indicate the order of priority and then set up a subgroup to work on the first item on the list, and so on.

Yes, we know. Some of the items will take some up-front money. So did the "superfragilistic bombardment mechanism." For capital funds the subgroup will have to pull together a case good enough to persuade management and perhaps the capital funds committee that the investment will pay off.

Use of the productivity index can help them make that case. If the improvements they propose increase the productivity index by, say, one point, the savings in development time, effort, cost, and reduced defects are worth about so much. This number can be calculated for your situation. We have seen a return on investment of more than 70 percent annually. That is generally enough to meet most companies' capital-investment hurdle rate.

Another kind of up-front money trains software people on the new tools or equipment that the capital investment acquires. Companies usually charge this money to various expense accounts, but they still have to cover the expenditure. One way to ensure that the funds will be available is to take advantage of the tradeoff savings that come from using smaller teams and taking a little more development time.

You're still not sure the first-priority improvement item is going to work? "The proof is in the pudding," your grandmother used to say? Of course it is. Here is where the productivity index comes in again. Take it periodically. See if the changes the task force is making are working. If the productivity index is improving, you will be emboldened to go ahead with more items on the priority list. Besides, improving process productivity will be freeing up funds that used to be sopped up by inefficiencies.

If the productivity index is not improving, go back to the drawing board. Get the task force (or a new task force) to come up with a better process improvement plan.

If it's so easy...

—why isn't everyone doing it? Well, we don't want to noise this around too much, but the fact is, it *isn't* easy. It is difficult to build good software. It is difficult to build a better process for building good software. But it can be done. It *is* being done, as our database demonstrates.

One thing that makes it hard is that managers sometimes have to pay a psychological price for getting out in front of the crowd. "The price for innovation is criticism, a price any creative spirit must be willing to pay," says Don Frey, once vice president of product development at Ford Motor Company and later chief executive of Bell & Howell [2].

Frey spearheaded disc brakes, radial-ply tires, unitized body construction, and other innovations at Ford. Years later Henry Ford II told David Halberstam (recounted in his book *The Reckoning*), "Frey may or may not be a genius but he is a pain in the ass!"

Fortunately for Frey, his innovations were successful in the marketplace. The market measured them and they met the test. In a parallel fashion, the productivity index can measure innovations in the software process.

Without innovation, organizations run down. The newspapers chronicle this sad fact every day: "Corporation X is laying off 1,000 managers."

Somebody ought to do something! Maybe it ought to be you. In most companies the software process alone is a potential gold mine. Getting it right strengthens the whole business.

> *"Clean up your own mess."* —Robert Fulghum, *All I Really Need to Know I Learned in Kindergarten* [3]

References

[1] T. DeMarco and T. Lister, *Peopleware: Productive Projects and Teams*, Dorset House Publishing Co., New York, N.Y., 1987, 188 pp.

[2] D. Frey, "Learning the Ropes: My Life as a Product Champion," *Harvard Business Rev.*, Sept.-Oct. 1991, pp. 46–56.

[3] R. Fulghum, *All I Really Need To Know I Learned In Kindergarten*, Ballentine Books, New York, N.Y., 1989, 196 pp.

"You've been up in the air long enough, Finlayson. You may come down if you promise not to float any more innovative ideas."

Chapter 33

Measure, Plan, Invest, Remeasure, Invest Again

"I'm familiar with improvement plans," the division president said. "In fact, we have several of them under way."

"Yes, but not in software development," we noted.

"That's true," he agreed. "I guess there are several reasons for that. For starters, software just sort of crept up on us. For a long time it was on the edge of the concerns uppermost in our minds."

"We keep making the point that software is strategic for most businesses," we said. "You have to have it."

"We've been coming to that realization for some years," he admitted. "For a long time, though, software was just a pain in the neck. There were all those long-haired guys running around in sandals doing something mysterious. We didn't know how to manage it or them.

"You can manage software," we offered helpfully.

"Yes, you've made that clear—plan, trade off, control, all those things," he allowed. "They make sense. It makes sense to 'do what you know you should.' As we move beyond the obvious, however, I think what to do becomes more obscure."

"We beg to differ," we submitted. "In broad terms, at least, we're sure you've been doing these things in other areas for decades."

We jotted them down on his blackboard:

1. Measure where you are now.

"That's one of the things the productivity index tells you," we interjected.

2. Get together a task force to develop your process improvement plan.

"Have them pick a few important items to work on first."

3. Measure where you get (a year or so later).

4. If your productivity index has improved, do some more.

"If not, rethink the plan and do something more promising," we said.

"The productivity index is a new way, at least for us, of measuring," he noted. *"Those steps we had in business school. What bothers me now is step 2, developing the process improvement plan. I'm afraid we are going to be snowed with weird-sounding proposals such as fourth-generation languages, graphical programming, and object-oriented programming. Who can keep up with all these brain twisters?"*

"Your software technologists, for one. The people who run short courses, for another. Consultants do this. Some experts write books."

The way to process improvement

In theory, resource-allocating levels of management favor process improvement. In manufacturing, making the case for a more efficient machine is straightforward. In software development, making the case for a specific improvement in the process is much more difficult. Then after making the case, it is hard to make the improvement. After making the improvement, it is hard to measure the fact that it is making a difference. Finally, it is hard to calculate a return on the investment from that measure.

Intermediate levels of management also favor process improvement, in theory. In practice, they have budgets and schedules to meet. They have fires to put out. In spite of slogans like "TGM," they may have little or no resources to work with. *

Raymond Dion at Raytheon obtained management approval for funding a process-improvement program by asking the managers themselves to become involved in identifying weaknesses and improvements. "We factored these ideas into the next briefing, gradually incorporating the suggestions of all key participants," he reported. "In this way, the program's ownership was spread throughout the division" [1].

---

* TGM, or Total Good Management, is a relic from the adventures of Phil and John. According to John, "Nothing much happened here in software development, except that we wore buttons that said 'TGM' for a while."

Below the management levels, software professionals would like to work more effectively and many of them do enhance their abilities over time. That is why experienced people are more efficient than inexperienced ones. In fact, after working on the development and introduction of the Capability Maturity Model for many years, its principal author, Watts S. Humphrey, became convinced that working software engineers had to personally experience the effectiveness of these methods before they would consistently apply them. To this end he developed the Personal Software Process [2].

Still, many of the roadblocks to a more efficient process are organizational in nature. They cannot be overcome by individuals working separately. They require organized attack. If the organization provides no means for this attack, the professionals withdraw into their shells. That is, they focus on the interesting problems in their immediate work and give up on the organization. The software process as a whole remains mired at its present level.

It is evident that software development organizations need a process to improve their process. They need a way to involve all the software professionals in this improvement process. They need a way to obtain the support of all levels of management and then to maintain that support.

A way to start

Some organizations have moved into process improvement through their efforts to prevent defects. If you are doing software development, you already have some kind of Process Stage, as diagrammed in Figure 33-1. You probably have two more of the figure's blocks, at least in rudimentary form: Inspection Team and Fix Defects.

Action team loop

Defects, in addition to being errors to correct in the current product, also represent flaws in the process that is making the product. People can establish a second feedback loop, the Action team loop, to fix process flaws. Soon after a defect is encountered, an analysis group studies it, looking for the root causes.

Tom Gilb and Dorothy Graham prefer to call this analysis brainstorming because it is carried out quickly by the same professionals who made up the inspection team [3]. Since they made the errors, they are in a good position to identify the causes. (In general, managers do not participate in this brainstorming meeting. Professionals usually feel freer to explore the causes of their errors without their merit reviewers looking on.)

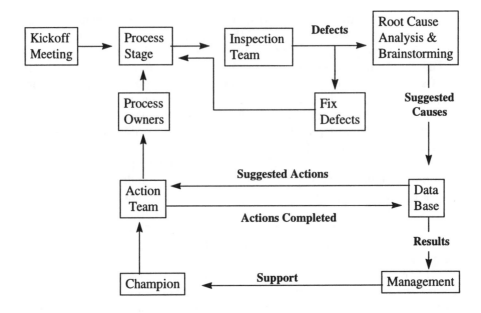

Figure 33-1. The first loop, through the Fix Defects block, fixes errors in software products. The second loop, through the Action Team, fixes flaws in the process. The third loop, through Management, supports and finances process improvement.

Robert B. Mays at IBM classifies errors into five cause categories: oversight (developer fails to consider all cases and conditions), lack of knowledge (achievable through education or training), communications failure (our old friend, ambiguity), transcription error (knows better), and process cause [4].

"If schedule pressure is the cause of oversights," Mays wrote in tracking one error category back to its source, "management should adjust their planning rates to allow more time in the schedule to do the work properly." Amen, we add.

The possible causes of defects then become the first entry to a database. As the causes are worked on, the database is updated.

The causes pass on to an Action team. This team has to have the capability of transforming possible causes of process flaws into action items. It needs members with special skills, such as overseeing process changes, coordinating training, and developing tools. It needs representatives from key functional areas, such as requirements, design, coding, and test. Moreover, it needs a management member to arrange for resources in addition to those of the team members themselves. This manager negotiates with other managers hours and schedule for people who are to add their efforts to those of the team.

Action items that affect the process itself pass to "process owners," the people or managers who have authority over a procedure, methodology, or tool. The process owners make the actual improvement.

The introduction of a new way of working into an organization often takes the intensity of a "champion." Management, of course, has those fires to attend to. The champion is a single individual who personifies the new way. Of course, he should know something about the technical content of process improvement, but his main assets are enthusiasm, persistence, and time to work on the process. While the champion's box has a specific location on Figure 33-1, he is in reality all over the place. Usually a champion emerges and is then sanctioned by management. It is hard to make a good one just by managerial fiat.

The process of process improvement began with an analysis of the root causes by working staff members. It should close with feedback to them of what is being done. Feedback may be newsletters, bulletin boards, staff meetings, written procedures, or training sessions. The psychology is that people who start the process have to see closure. The fact that something is being done about a set of causes provides the motivation for people to generate the next set of causes. Going around this feedback loop is not a one-time journey.

Meantime the participants in the improvement process are recording what happens in the database. The database provides a means of tracking and follow-up, so that action items don't fall through cracks. It records metrics from which results, such as improvement in the process productivity index, can be derived that inform management that the program is worth what it costs. The metrics also tell the full circle of participants that the improvements are working.

We may seem to have implied that Action teams work only on actions derived from individual defects. That is true, of course, but the teams also aggregate defect causes into patterns where an action item may correct many similar causes. A training program, for example, is going to deal with a pattern of problems, not just one defect cause.

Moreover, the action team provides a focal point to which a software professional can direct an improvement idea or where management can initiate a major improvement. The fact that some level of improvement activity is going on all the time (and some level of results is being achieved) leads both software people and management to become comfortable with the process of improvement.

Management support loop

"It was not clear at the beginning of our work on the Defect Prevention Process that software defects could be prevented," Mays wrote. "However, our experience with this process has shown that not only are defects preventable, but significant reduction in errors can be achieved with a modest investment. Software defects have identifiable causes, such as an oversight or communications failure, and are preventable through improved processes, methodologies, techniques, and tools. A dramatic improvement in quality can be achieved through defect prevention and with it a corresponding improvement in overall productivity and customer satisfaction" [4].

Mays cited cost figures for IBM projects employing this defect-prevention process ranging from 0.5 percent to 1.3 percent of the projects' costs. These costs are spread over defect-prevention meetings, 16 percent; action implementation, 60 percent; and education classes, 24 percent.

"An investment of about one percent of an area's resources in the Defect Prevention Process will return somewhere between six and eight times the cost as savings to the organization," Mays said. "A typical organization of 200 people, which spends about 1.8 person years per year in defect prevention activities and achieves a 55 percent reduction in defects, will realize a savings of between 11 and 15 person years per year" [5].

He went on to observe that "the most significant benefit of the Defect Prevention Process is higher product quality in the field."

Computer Sciences Corporation invested about 1.5 percent of its project budget over a period of two years to get a similar reduction in error rates [6].

Data like these, cited not only by Mays, but also by Carole Jones [7] and Gilb and Graham [3], help to overcome one of the problems referred to above—getting the commitment of top management to an improvement process. Besides, this process can start small, with just one causal analysis meeting and one action team, if necessary. The number of people devoted to the effort can grow as the results show the process is working.

"That seems to cover all the bases," Phil observed. "It starts where you are; it involves everybody; it encourages everybody with feedback; it goes on continuously—you don't run out of ideas to work on until you run out of errors; and it gives management the results that keeps the process going around the loop."

"When we run out of defects," John proclaimed, "that will be the day."

References

[1] R. Dion, "Elements of a Process-Improvement Program," *IEEE Software*, July 1992, pp. 83–85.

[2] W.S. Humphrey, *A Discipline for Software Engineering*, Addison-Wesley Publishing Co., Reading, Mass., 1995, 789 pp.

[3] T. Gilb and D. Graham, *Software Inspections*, Addison-Wesley Publishing Co., Reading, Mass. 1993, 471 pp.

[4] R.G. Mays et al., "Experiences with Defect Prevention," *IBM Systems J.*, Vol. 29, No. 1, 1990, pp. 4–32.

[5] R.B. Mays, "Defect Prevention and Total Quality Management," in *Total Quality Management for Software*, edited by G. Gordon Schulmeyer and James I. McManus, Van Nostrand Reinhold, New York, N.Y., 1992, 497 pp.

[6] D.N. Card, "Defect Causal Analysis Drives Down Error Rates," *IEEE Software*, July 1993, pp. 98–99.

[7] C.L. Jones, "A Process-Integrated Approach to Defect Prevention," *IBM Systems J.*, Vol. 24, No. 2, 1985, pp. 150–167.

"One more question. Do you know anything about software process improvement?"

Chapter 34

Simplify the Project

"In a sense, creating software requirements is like hiking in a gradually lifting fog. At first only the surroundings within a few feet of the path are visible, but as the fog lifts, more and more of the terrain can be seen."
—Capers Jones [1]

A key factor in software development is the system requirements. It is key because the requirements largely lay down the eventual size of the product. They establish to which application type the system belongs. Their intricacy affects how difficult it will be for a project team to specify, design, and build the software. The application type and design difficulty, in turn, influence where on the process productivity scale the work will be done. These two, size and process productivity, are factors in the software equation that affect the amount of effort and development time required.

When you reduce the expected size of a project, the remaining management numbers—effort, schedule, and defect rate—all improve. To the extent that you can simplify project requirements, you can reduce product size. The data on historical projects, presented in Figures 4-1, 4-2, and 4-3, clearly shows that at smaller sizes schedule time declines, effort declines, and defects decline. Similarly, the software equation (where Size equals Process Productivity times Effort^1/3 times Time^4/3) demonstrates that, Process Productivity holding constant, the product of Effort and Time must decline when Size decreases.

It follows, then, that decreasing size leads to reduced schedule time and effort. The issue becomes how to decrease size and one approach is to simplify the requirements.

Simplify requirements

Customer organizations tend to fancy feature-loaded systems. They ask for features not central to the product's main functions. They urge features that are nice to have. They even add features that only one or two especially knowledgeable users are likely to use. In short, they adore "bells and whistles." It is not that any one person in the customer organization wants all these addenda. It is that the fancies of many individuals add up to a large, complex product.

Some managers welcome this growth. As they see it, large requirements turn into large projects, and large projects lead to funds, staff, and development time. Unfortunately, the customer may have an appetite larger than its purse. It may need the product sooner than the expanded requirements permit. Further, large projects may exceed the capacity of the available project people, leading to failure via another path.

Worse, the multiplication of features leads to additional interactions between them. This complexity means the project takes longer to produce. For example, the US Army awarded a contract worth $34 million for the Advanced Field Artillery Tactical Data System to Magnavox Electronic Systems Co. in May 1984 with delivery set for 1987. The system was to automate an Army battlefield. Neither the Army nor Magnavox had thought through all the details of what this effort involved, although they did have 2,000 pages of requirements. Naturally, development time and costs escalated [2].

Magnavox completed the initial contract in 1989, two years late. The Army had capped the cost at $46 million, but Magnavox contributed another $30 million of corporate funds. So the cost had more than doubled. Meantime the Army had redefined the project as a "concept evaluation." Magnavox now received a new $60-million contract and another three years to develop the next phase.

Doing something like automating a battlefield for the first time is evidently more difficult than the Army anticipated. One of the generals overseeing the program did derive an important lesson from the experience.

"With large programs, it's a very difficult thing to get it all working and out there in one fell swoop," said Major General Peter Kind. "It just doesn't work that way."

For a system as novel and complex as an automated battlefield, some difficulty with requirements is not surprising. "Where a new system concept or new technology is used, one has to build a system to throw away, for even the best planning is not so omniscient as to get it right the first time," Fred Brooks noted as far back as 1975 [3].

In fact, getting requirements right is the central issue that Donald C. Gause and Gerald M. Weinberg explore [4].

Getting requirements wrong, or partially wrong, leads to certain project failure. It is often the reason behind the widely publicized disasters that hit the newspapers. For projects exceeding 10,000 function points (about 1,000,000 source lines of code), Capers Jones estimates the probability of failure—of cancellation of the project—to be 65 percent [5]. The probability declines to less than 25 percent for projects under 1,000 function points. The cost to the US economy was about $14 billion in 1993, he estimated, or about 285,000 personyears of effort. That is about 15 percent of the total effort devoted to software development.

Of course, there are other reasons for failure besides poor requirements. For instance, "there are no commercial or proprietary estimating tools that can accurately predict the follies of incompetent management or personnel or the demands of unreasonable clients, and the chaos which they may create," Jones observed.

Develop incrementally

Requirements analysts cannot list specifications for features that no one has even imagined yet. As users discover during system development and test features that they had not initially imagined, the product grows in size and complexity. The general answer to incomplete initial requirements is incremental development.

Rapid prototyping [6], joint application design [7], evolutionary development, spiral development [8], iterative design refinement [9], and periodic release are terms that people have applied to the process of initially specifying only the central features of a proposed system. This initial system typically performs only the main-line functions, excluding the peripheral functions that add to complexity. The software engineers develop this relatively simple system. Then they try out this prototype, possibly in house, or at beta sites, or in a preliminary release.

With the knowledge gained in the first go-around, they specify a more extensive product. It probably will be more complex than the first project, but now everyone has more experience. With this additional knowledge, people are better able to cope with the added size and complexity. Revised estimates of effort and schedule can be more accurate.

"The effort required for evolution of a software system can be reduced through prototyping," Luqi wrote in a theme issue of *Computer* devoted to rapid prototyping. "Prototyping can stabilize the requirements for both new systems and proposed enhancements to existing systems" [6].

Ivar Jacobson and his co-authors contend that models are necessary to communicate business ways both to the stakeholders, including the process users, and the software developers [10]. He describes the first step in the modeling process: the preparation of use-case models. The method begins with analyzing what customers want to get from the company. These "uses" are called use cases. They are reduced to formal models that form the initial basis for the reengineering effort.

Of course, many other "methodologists" have considered this and comparable processes—real-time and embedded systems, for example. "A problem with past system specification methods," write Derek J. Hatley and Imtiaz A. Pirbhai about real-time systems, "has been that they tend to address only one aspect of the system, whereas systems actually have many aspects" [11].

Merlin Dorfman and Richard H. Thayer have collected scores of papers on systems and software requirements engineering in two volumes [12] [13].

Discipline change

Change requests from the customer or marketing once a project is under way usually have serious effects. For example, a software engineer in a training program we were conducting in a large company reported a case to us that hit our funnybone. One of the company's marketing managers had gone to a trade show earlier that week. There he had seen a PBX switching product in a cabinet he found far more attractive than the cabinet for his company's corresponding product. The competitor's cabinet was six inches shorter! His own cabinet was just about to go into production.

"We've got to modify our cabinet," the marketing manager told the engineering group. "Their cabinet is much better looking. I'm sure it is going to sell very much better than ours. Cut six inches off our cabinet."

The engineers looked inside their cabinet. It was stuffed with circuit boards, backplanes, power supplies, connectors. It might be possible to take six inches off the height, but it would mean redesigning the boards and backplanes, even revising the ROM (Read Only Memory) software.

We left the plant the next day, so we don't know what the company finally did. We trust that it had a change-control procedure that forced all concerned to take a hard look at each proposed change. Responsible executives should weigh additional design costs, schedule delay, availability of engineers, factory costs for delaying production, and so on. On the marketing side, maybe the smaller cabinet would sell better. How much better would it have to sell to cover those additional costs?

With a hardware product, one can see the complications. With software, they are less obvious. It seems simple to change or add a few lines of code in contrast to the complications of redesigning and remanufacturing a physical part. Often it is not simple. The change may reverberate throughout the system.

The reason is that changes to software often increase complexity. In addition, changes usually increase the number of lines of code, leading to increased effort and schedule time. Therefore, both to reduce complexity and inhibit size growth, organizations need a formal change process to assure that informed people fully analyze these complications. Think through the implications of each potential increase in complexity or size.

Change control is one of the core operations in configuration management. Other operations include the identification of the items to be controlled and the establishment of a baseline version [14].

Make subsystems independent

A major contributor to the difficulty of designing and building systems is sheer size. The interactions between many different parts of a large system increase complexity. If the system engineers could separate a system into several entirely independent systems, then each new system would be less complex than the original very large one. Designers could develop each independent system separately in less time and with reduced effort.

Creating entirely independent systems is usually not possible because there are interrelationships within the overall system. The advantages of independent systems point to the desirability of reducing and simplifying the relationships that are present between the subsystems. During the early stages, systems engineers should define these relationships carefully and completely. In this way the subsystems, though still part of the larger system, can be effectively less complex.

> *"I took a night course in programming about 20 years ago when I was still down in the ranks," the vice president of finance told us. "At first all those instructions were confusing. It was like learning a new language. By the end of the semester, however, programming seemed simple enough."*

> *"Perhaps 'simple' is not quite the right word," we said. "Learning language is tough for most people, but the complexity of software is something else. Fred Brooks, for example, believes that complexity is the essence of a software entity, not an accidental property" [15].*

> *"What does he mean, accidental property?"*

> *"Accidental refers to those difficulties that attend software development with today's methods, but are not inherent in the process," we responded. "We could imagine a design method, for example, that would provide a design description in such a precise way that a computer program could generate code from it."*

"I can imagine that, but I can't imagine a computer program that would generate requirements from our business situation," the vice president said sharply.

"Exactly. That's essence," we said. "That difficulty is inherent in going from the informal real world to the precise world of the computer."

"That is where we businessmen live—in all that blur," he replied. "I guess the programmers live over there in that more precise world."

"Yes, once you get the business problem to that stage, the mechanics of coding it are not very complex," we agreed. "Training and experience in software coding are not of themselves a great help in dealing with the uncertainties of users' needs and desires."

"You seem to be implying that those of us on the business side need to contribute a little more to this process of going from the informal to the precise," the vice president suggested.

"Yes, you people have more experience in the real world. If the product that finally gets to the user has unneeded features, or if it is less reliable than a simpler product would be, that affects their 'delight' in it," we said. "That is a matter that affects the company and you."

"No doubt of that," he agreed [15].

References

[1] C. Jones, "Strategies for Managing Requirements Creep," *Computer,* June 1996, pp. 92–94.

[2] E. Williams, "The Software Snarl," *Washington Post,* Dec. 9–12, 1990.

[3] F.P. Brooks Jr., *The Mythical Man-Month: Essays on Software Engineering,* Addison-Wesley Publishing Co., Reading, Mass., 1974, 195 pp.

[4] D.C. Gause and G.M. Weinberg, *Exploring Requirements: Quality Before Design,* Dorset House Publishing, New York, N.Y., 1989, 320 pp.

[5] C. Jones, *Assessment and Control of Software Risks,* Prentice Hall Inc., Upper Saddle River, N.J., 1993.

[6] Luqi, "Software Evolution through Rapid Prototyping," *Computer,* May 1989, pp. 13–25.

[7] J. August, *Joint Application Design,* Yourdon Press, Prentice Hall, Inc., Upper Saddle River, N.J., 1991.

[8] B.W. Boehm, "A Spiral Model of Software Development and Enhancement," *Computer,* May 1988, pp. 61–72.

[9] P.G. Bassett, *Framing Software Reuse: Lessons from the Real World,* Yourdon Press, Prentice Hall Inc., Upper Saddle River, N.J., 1996, 384 pp.

[10] I. Jacobson, M. Ericsson, and A. Jacobson, *The Object Advantage: Business Process Reengineering with Technology,* Addison-Wesley Publishing Co., Reading, Mass., 1995, 347 pp.

[11] D.J. Hatley and I.A. Pirbhai, *Strategies for Real-Time System Specification,* Dorset House Publishing, New York, N.Y., 1987, 412 pp.

[12] R.H. Thayer and M. Dorfman, *System and Software Requirements Engineering,* IEEE Computer Society Press, Los Alamitos, Calif., 1990, 735 pp.

[13] M. Dorfman and R.H. Thayer, *Standards, Guidelines, and Examples on System and Software Requirements Engineering,* IEEE Computer Society Press, Los Alamitos, Calif., 1990, 620 pp.

[14] F.J. Buckley, *Implementing Configuration Management: Hardware, Software, and Firmware,* IEEE Press, Piscataway, N.J., 1992, 256 pp.

[15] F.P. Brooks, Jr., "No Silver Bullet: Essence and Accidents of Software Engineering," *Computer,* Apr. 1987, pp. 10–19.

Chapter 35

The Process of Reusing Software

"Let's dedicate our efforts to educating those searching for 'silver bullets' that reuse is not simple, but has tremendous potential for changing the way we develop and maintain software—if we systematically attack the technical, contractual, legal, management, and social issues associated with reuse." —Donald J. Reifer [1]

One of the most efficient ways to develop software is not to develop it at all! Reuse an existing program instead. If you cannot reuse an entire program, reuse parts of one. Since there is more to software development than code, reuse your architecture, specifications, designs, test plans, documents, or parts of them.

Reuse increases code-production rate

Reusing these components, in effect, reduces the size of the proposed system. Reusing some components reduces the number that have to be developed from scratch. There is some cost of reuse, of course. Reducing the size can shorten development time and reduce effort and cost.

QSM Associates (Ira Grossman and Michael C. Mah) demonstrated these effects in a study of nine organizations in 1994. Two of the organizations prefer to remain anonymous, but the other seven are large and well known:

Ameritech
Chemical Bank
Hudson's Bay Company
Noma (Cable Technology)

Revenue Canada

Teleglobe Insurance Systems

Union Gas Limited

The nine organizations contributed data on 15 projects, all in the business-systems application area. Within that area, however, the projects included customer tracking, sales and financial analysis, data interchange, retail, training, and other business functions.

The common denominator of all the projects was the use of frame technology. This technology was invented by Paul G. Bassett between 1978 and 1981 and has been marketed since then by Netron, Inc. of Toronto, Canada. Bassett's technology is a way of packaging software into relatively large "frames"—software components that can be adapted to new circumstances and reused [2]. The purpose of the study was to find out how well frame technology accomplishes these purposes. Our more general intention here is to see what we can expect in the way of process improvement from an advanced form of reuse technology.

The result of the study, shown in Figure 35-1, was that all the projects produced source lines of code at a faster rate than the industry average rate. Seven of the projects exceeded one standard deviation above the mean, doing better than 84 percent of the projects in the QSM database. The largest project in the sample, at 9,300,715 SLOC, at one standard deviation above the mean, is consistent with the other projects.

The size count represents all the code turned out—new and reused. On average, 8.4 instructions were delivered for each instruction that was written new. That is a reuse ratio of 88 percent. The smaller projects—less than 60,000 lines of code—averaged 85 percent. The projects exceeding 100,000 lines of code averaged 90 percent.

Grossman and Mah noted, to their surprise, that "the organizations categorized as first-time users of the technology had the highest ratio of delivered instructions per written instruction." Netron supplies several hundred frames covering widely used business functions. Then, over time, the client develops division-wide and corporate-wide frames for its own repetitive functions that increase the proportion of reuse further. On looking into these first-time users, Grossman and Mah found the answer: Netron's experienced consultants had provided frame-engineering support.

Figure 35-2 presents the 15 projects in terms of conventional productivity: source lines of code per personmonth. All but two were well above average, mostly better than one standard deviation above the mean. On looking into the two below-average projects—both very large—the two consultants found that they had employed very large teams in response to high schedule pressure. As we have noted, that combination leads to greater effort and lower productivity.

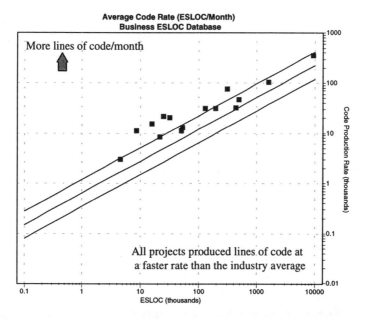

Figure 35-1. Projects employing frame technology, that is, substantial reuse, produced much more code per month, new and reused, than the mean trend line of QSM's business database.

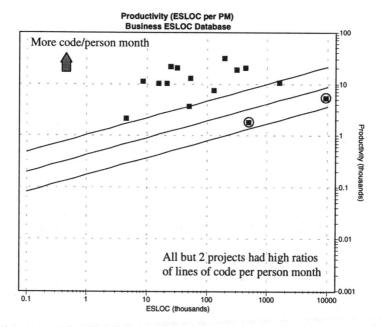

Figure 35-2. Projects employing reusable frames generally enjoyed high conventional productivity, but time pressure pushed two of the larger projects below average.

The wide dispersion of these conventional-productivity data points re-emphasizes the point that other factors besides the technology employed influence software productivity. One such factor, for instance, is the time pressure under which the project is carried out, as the two low-ranking data points demonstrate. This dispersion reinforces our view that conventional productivity is an unreliable metric to use in estimation.

Reuse improves management numbers

Extensive reuse, such as these 15 projects enjoyed, has a profound effect on the management numbers:

(1) *Schedule*. Cycle time was reduced 70 percent from the QSM database average. To put it another way, an average-size project in this sample (133,000 SLOC) took 5.3 months, versus our database norm of 18 months.

(2) *Effort*. Personmonths were reduced 84 percent from the QSM database average. The average-size project expended 27.9 personmonths, versus our norm of 179.5 personmonths.

(3) *Cost*. Translated into US dollars, the project savings would be on the order of one and a half million dollars, depending on the labor rate and overhead values used.

(4) *Defects*. Seven of the 15 projects had error data collected from the start of system testing to full operational capability. Four projects were near average on this metric; two were nearly one standard deviation better than average; one was one standard deviation worse than average. On looking into the details, Grossman and Mah found that this project had trouble during requirements definition and learning problems later.

(5) *Process productivity*. The average productivity index of the 15 projects was 26.2. That is 9.3 index points higher than the current mean of business applications: 16.9. Since the PI index scale stands in for the nonlinear process-productivity parameter scale, the process productivity of these projects was actually 9.4 times greater than average.

Reuse sounds like a beautifully simple answer to improving the management numbers but, as with many great ideas, it has problems. When you buy a spreadsheet program from Lotus, for instance, you are reusing a program. But Lotus has had to communicate to you through advertising, direct mail, or word of mouth the niche its program fills. It has had to provide you with a manual that tells you how to use its program. It has had to spread short courses throughout the world to train those who hate manuals. It has had to set up a wholesale and retail distribution system to bring the disks to your neighborhood. It has had to place an economic value, namely a price, on all this activity and persuade you to pay it.

If you want to reuse software components developed within your organization, you have to arrange something comparable, though on a smaller scale. Because setting up a library of reusable components requires investment, someone down in the ranks cannot casually take the initiative.

Costs of reuse

We can divide the cost of reuse into two categories. One category is the cost of preparing potentially reusable specifications, designs, modules, and documentation and getting them into some kind of library. The other is the cost of getting a reusable component out of the library and fitting it into a new system.

The first category includes a long list of possible costs, such as:

- Setting up standards to which to design potentially reusable components so that they will fit easily into later systems;
- Designing to these standards, as compared to designing just for the project at hand;
- Documenting a reusable component so that someone not on the original project can figure out how to use it;
- Establishing a library of reusable components (such as space, equipment, librarian, or a search system of some kind).

Some of these costs are capital expenditures. Others are current operating expenses. In neither case will the project on which the component was originally developed happily bear the additional cost. In a broad sense these costs are overhead, chargeable over a number of present and future projects. In any event they are something that management has to figure out how to handle.

In addition to these costs there are other problems. For instance, how do you motivate the people on a rush project (they are all rush projects!) to spend extra time preparing reusable designs and modules? How do you motivate people on later projects to "not invent it here?"

The second category of cost of reuse is chargeable to the project making use of the reusable component. Reuse is not cost-free. These costs include:

- The cost of a designer searching for some component in the library and deciding that it is, indeed, applicable to the project at hand;
- The cost of figuring out that no modifications are necessary; or
- The cost of making some modifications;
- The cost of integrating and testing an unmodified or modified module to assure that it functions in the new setting.

The cost of using an unmodified module may be on the order of 15 percent of the cost of designing and coding an entirely new module. Depending on the degree of modification, the cost of a reused component runs up to 100 percent of the cost of new design. Of course, beyond 100 percent (if you can figure that out beforehand), it is cheaper to design anew.

Thus, the cost benefits of reusing components is, in reality, greatly reduced by the investment cost of creating, standardizing, certifying, and warehousing potentially reusable components and by the cost to a project of incorporating them into a new program. For example, Toshiba's studies, as reported by Michael A. Cusumano, indicated that productivity improved significantly only if about 80 percent of a module or more was reused without changes [3]. If less than 20 percent was used unchanged, the impact on line-of-code productivity was negative. Between 20 percent and 80 percent reuse, the impact on productivity was not noticeable.

The four big Japanese software factories studied by Cusumano have considerable experience in reuse, having been systematically reusing components since the mid-1970s. For example, Toshiba increased lines of delivered code taken from existing software from 13 percent in 1979 to 48 percent in 1985. It increased code production from 1390 source lines per person per month in 1976 to more than 3100 in 1985. Of course, there were many reasons for this improvement, but one was reuse.

Systematic attention to standardizing and certifying software components could move more of them to the point where the impact on productivity is substantial. That implies incorporating reuse strategies into the software process.

Strategies of reuse

Some researchers in this field question whether the technology is sufficiently developed to be widely applied and, of course, it is true that much work is still underway [4]. Nevertheless, the several dozen companies sending representatives to the annual Workshop on Software Reuse are using reuse methods to various degrees.

Some organizations are attempting to obtain the benefits of reuse by simply ordering it into existence. However, the preferred tactic seems to be to *grow* it under the inspiration of local champions or change agents, but with support from higher levels of management.

There is also a spectrum of reuse approaches from the informal to the highly organized. At the informal end is the developer who reuses some of his or her own code or that of coworkers nearby. That approach takes little management attention. This "passive" approach, according to Ed Yourdon, "typically hovers at the 15 to 20 percent level in most IT organizations today" [5].

As the possibility of reuse spreads beyond a few close colleagues, a formal structure becomes needed to identify code or other reusable software components. Still more structure is necessary to identify components that originated in many different organizations.

Components may be divided into three levels. The first level has been long familiar—code skeletons, macros, copybooks, and subroutines. Each requires a programmer to copy a master, modify it, insert it, and test it. In the process it loses its identify as a reusable component. The output of code generators is a form of this reuse.

The second level is object-oriented technology. An object may inherit capabilities from another object. In effect, it is reusing that capability. But there are drawbacks. One is that individual objects are small; there tend to be thousands of them. Consequently, programmers find it difficult to keep track of what is available. They may find it easier to recreate a needed function than to search for it.

The third level is larger components: frames, frameworks, and patterns. These sets of components number in the few hundreds. They greatly reduce the search and understanding load on the user.

This spectrum of levels seems to imply that people create software much as usual and somebody sorts out what may be reusable and makes it available somehow. One step further is to figure out first what is reusable and to design it in such a way that it is, in fact, readily reusable. The name, domain analysis, has been applied to this approach.

Hewlett Packard's Queensbury Telecommunications Division in Edinburgh, Scotland, has been applying this approach to instrument development.

"We looked at the source code of instruments that had been built in the past to try to identify areas where generic solutions could be applied," Malcolm Rix reported to the Fifth Annual Workshop on Software Reuse (1992). "Code that looked very systematic, exhibited a high degree of regularity, or was similar from one product to the next, was identified as targets for the reuse system."

In the case of this instrument domain, the reuse level is now at about 60 percent (by code lines). Rix estimated that about 20 percent of engineering time had been saved, compared to running projects independently.

The possibilities

Shrink-wrapped software products now sell to millions of ultimate consumers for a few cents per source line of code. That cost probably represents the ultimate in reuse.

Next in line are the outsource contractors. To a considerable extent, they are still operating their client's existing code. As they substitute their own programs in scores or hundreds of clients' computer rooms, they will be reusing on a giant scale and greatly reducing software development costs.

In contrast, to develop a line of new code from scratch in the traditional way costs in the range of dollars, not pennies. The cost and time-to-market impetus to move up the reuse ladder will be a hard driver.

References

[1] D.J. Reifer, "Software Reuse: The Next Silver Bullet?" *American Programmer* (theme issue on Reuse), Aug. 1993, pp. 2–9.

[2] P.G. Bassett, *Framing Software Reuse: Lessons from the Real World*, Prentice Hall Inc., Upper Saddle River, N.J., 1996, 384 pp.

[3] M.A. Cusumano, *Japan's Software Factories: A Challenge to U.S. Management*, Oxford University Press, New York, N.Y., 1991, 513 pp.

[4] W. Myers, "Workshop Participants Take the Pulse on Reuse," *IEEE Software*, Jan. 1993, pp. 116–117.

[5] Yourdon, Rise & Resurrection of the American Programmer, Yourdon Press, Prentice Hall, Upper Saddle River, N.J., 1996, 318 pp.

"Of course you have good ideas, dear. It's just that you don't *structure* them."

Chapter 36

The Process of Business Reengineering

"[Information technologies] have become the principal means of improving existing business processes." —Paul A. Strassmann [1]

Before the industrial revolution a shoemaker, perhaps with a few apprentices, brought in hides, cut leather, sewed shoes, and sold them out the front door to customers who put them on and used them, as diagrammed in Figure 36-1A. The shoemaker coordinated the process, kept the information for doing so largely in his head, made the shoes, and sold them. In modern terminology he was the whole value chain.

This pattern evolved into Figure 36-1B. A master shoemaker coordinated the work of a number of journeymen and apprentices. He began to keep some information on paper to aid his memory. The work was divided into jobs. Disposal of the product became a little more complicated, at first to a nearby shoe store and later to many stores.

Coming down to recent times, before the advent of computerized information systems, the pattern of production and service looked something like Figure 36-1C. The value chain is still there, between suppliers and buyers, but it has become longer and more complicated. The receiving operation, for instance, may involve matching the incoming supplies against a purchase order, inspecting the material or parts received, rejecting defective parts, warehousing accepted parts, issuing them to factory operations, and so on. In a modern factory or service operation there may be scores of occupations. This is division of labor, as envisioned by Adam Smith several hundred years ago.

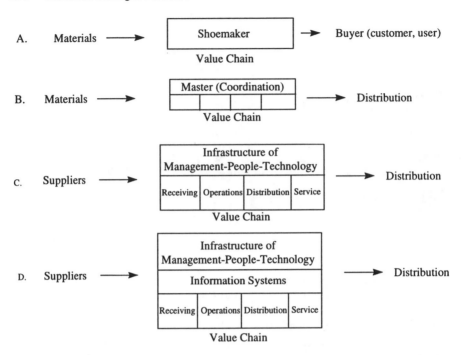

Figure 36-1. (A) In older times a simple shoemaker coordinated his own value chain. (B) A little later, a master shoemaker coordinated the work of a few journey-men and apprentices, keeping the details in his head. (C) As manufacturing and services grew more complex, a substantial administrative infrastructure developed to coordinate the many operations. (D) With the advent of the computer, organizers could transfer part of the coordination load to Information Systems.

Along with division of labor, however, came the necessity to coordinate this growing array of occupations, machines, materials, and technology. A vast infrastructure of management and administration grew up to acquire, keep track of, and apply a great deal of data to this task. Because the quantity of this data exceeded the ability of a single supervisor to sift and utilize it, a hierarchy of managers came into being. This hierarchy was assisted by an even vaster army of clerks for material control, production control, quality control, accounting control, and so on. The efficiency gained from division of labor in the value chain was supposed to offset the increasing cost of the infrastructure needed to coordinate it.

Much of this cost was labeled "overhead," and all right-thinking managers strove mightily to reduce it. That was a worthy objective, but coordinating all those operations did take a considerable infrastructure. If the coordination broke down and the value chain hiccuped, the cost was even worse.

Coordination is based upon information. Most of this information comes from the processing of large quantities of data into forms that enable supervisors and managers to act. Since this processing was done by employees in the period epitomized by Figure 36-1C, the figure does not show information as a separate block. Moreover, the information collected during this period was not always complete or accurate. Often the resulting data failed to meet all the needs of managers. We have all seen a harassed manager, busy with telephone calls, pestered by subordinates, seeing for himself by walking around. To the extent that the information system is complete, however, there is less need to walk around to collect information informally. (It may still be desirable for other purposes, such as gauging the morale of the people.)

As punch-card machines, then computers, and now information technology have developed, more and more of the data processing load could be performed by machines. A separate block, Information Systems, became appropriate, as shown in Figure 36-1D.

Focus on value

The first step toward the establishment of a computerized Information Systems block is to figure out what the value chain is. What is the chain of processes that Information is to coordinate? Sometimes management is preoccupied, not with the value chain, but with the operation of the infrastructure itself. Few process activities are as fascinating as office politics. At these times management may pay little attention to the value chain that is the ostensible object of the infrastructure's activity. In other situations an existing business may have grown by taking advantage of opportunities over the years with the result that the current conglomeration of functions doesn't make much sense. It may even be sliding into unprofitability.

Michael E. Porter, professor at the Harvard Business School, considered the value chain from two standpoints. What is the attractiveness of the industry? What determines a company's competitive position? [2] C.K. Prahalad of the University of Michigan and Gary Hamel of the London Business School approached the issue from another direction. What are the core competencies that the overall business brings to the support of a particular value chain? [3]

Michael Hammer and James Champy explained that establishing a new business process means "abandoning long-established procedures and looking afresh at the work required to create a company's products or service and deliver value to the customer." Ford Motor Company's reengineering of its accounts payable process reduced the number of people by 75 percent [4].

In the old way the purchasing department sent a purchase order to a vendor and a copy to accounts payable. When the goods arrived, material control sent a copy of the receiving document to accounts payable. A few days later the vendor sent an invoice to accounts payable. Accounts payable then checked 14 items across these three documents. Usually clerks spent considerable time reconciling the discrepancies. Finally accounts payable authorized the issuance of a check to the vendor.

In the reengineered way, the purchasing department sends a purchase order to the vendor, but enters the copy in an on-line database accessible to all the concerned functions. When the goods arrive, material control checks them against the on-line purchase order. If the goods correspond to the purchase order, material control notes that fact in the database. The computer system matches three items of the purchase order and the receiving record. If they match, it authorizes payment. Vendors have been instructed not to send invoices.

Establishing a series of processes from a value-chain point of view requires people to look at the operations of the business with the buyer's interest in mind. The "buyer" may be a customer or user, external or internal. In the case of a process entirely within an internal value chain, the "buyer" is the person at the beginning of the next process.

"Information technology offers many options for reorganizing work," Hammer has found. "But our imaginations must guide our decisions about technology—not the other way around."

Information up and down

The Information Systems block serves both the managers in the infrastructure above it and the people in the processes below it. For managers it should provide the various groups in management with the specific information each group needs for its particular purposes. Since the Information Systems do the data collection and analysis, the small army of employees that once performed these functions is no longer needed, or at least not needed in the abundance with which they were once present.

Similarly, since the information can be adapted by the System to the needs of each level of the management hierarchy, there is less need for the screening and condensing functions that lower levels of the management hierarchy once performed for the higher levels. When Information Systems are fully worked out, the management structure can be flatter. Flatter structures have led to thousands of middle managers being laid off. (In some cases, observers suspect that top executives have flattened the structure and laid off middle managers without actually carrying out the reengineering process. They expect—at least subliminally—the remaining managers to work longer and harder. Of course, there is a limit to "longer and harder," and some companies are running up against it.)

For the people in the value chain, the Information System enables them to keep their work synchronized with groups before and after them in the value chain. They no longer depend on coordination by several layers of supervisors and managers.

When all this is worked out successfully, the people become "empowered." That is, they have the information needed to serve the buyer at the end of the value chain successfully. In this sense, the sense of coordination through data, you can't empower people with slogans, wall posters, or inspirational speeches in the company cafeteria. At the same time, however, you don't turn a couch potato into an empowered value adder without some effort in the way of education, training, and motivation. Moreover, the remaining supervisors and managers sometimes find it difficult to refrain from giving orders in the old way. Both managers and people have access to the same information from the System; both know what they have to do to satisfy the buyer at the end of their process chain.

The business processes of the first decades of the computer era were those necessary to operate Adam Smith's division-of-labor concept. In Smith's day payroll clerks with quill pens made up the payroll. Early programmers committed the successors to these quill-pen procedures to programs to run on mainframes. It was so for business processes in general. Very soon, however, managers and developers noticed that they could improve existing procedures. They made incremental improvements, but the procedures were basically traditional.

Information enables reengineering

In the last decade, farsighted people began to see that information technology could be the means to dramatically overhaul the existing ways themselves. Computerization could do more than just incrementally improve the old ways.

In the traditional case—and there is still plenty of this work going on—the existing business procedures are the source from which analysts draw the requirements. These procedures had been around for a long time and were fairly well fixed. It was possible to derive firm requirements from them. Given good requirements, engineers could follow the waterfall model of system development. That is, they could proceed through specification, high-level design, detailed design, implementation, unit test, and system test in sequence.

In the course of figuring out what the existing procedures were, both users and analysts noticed ways to improve them. Sometimes they noticed the improvements in time to get them into the initial requirements.

Much of the time the improvements grew out of the process of development itself. As users and developers worked on the system, they found improvements. Accepting improvements implied a change process, configuration management, revised estimates, and often a change to design, specifications, or requirements back upstream.

In the new approach, business process reengineering, the existing way is being totally replaced by a new way. The complete requirements for the new way cannot be wholly known until it is fully developed. But it cannot be fully developed until the Information System, which constitutes a large part of the new way, is worked out. We have a "chicken or egg" situation.

We don't know what the chicken ought to do, but software developers in business reengineering situations have to work incrementally. That is, from a sketchy initial set of requirements they develop a first iteration that enables the business reengineers to grasp in general what the Information System is going to do for them. From this sense they are able to work out requirements in further detail. From these specifications the developers can produce the second iteration. Iteration continues until the Information System becomes useful to management and people, and the developers issue the first full-fledged release.

If each one of these iterations were to be a two-year project, reengineering a business process would take four, six, or eight years. One would like to be able to turn around each iteration in months, rather than years. The only approach that can speed up development to that degree is reuse. However, reuse can work that fast only if the reusable components are fairly large. We referred to one such pattern of reuse, Paul Bassett's frame technology, in Chapter 35.

Object technology targets reuse

Proponents of object-oriented technology have long advocated its application to reuse. The mechanism of inheritance, contained in object-oriented languages, provides a means. A number of development methodologies based on object-oriented technology have been evolving:

Grady Booch, Booch method

Peter Coad, Coad method

Ivar Jacobson, Objectory method

Ralph E. Johnson, Frameworks

Stephen J. Mellor and Sally Shlaer, Shlaer-Mellor method

James Rumbaugh, OMT method

Rebecca Wirts-Brock, Responsibility-driven design method

Booch, Jacobson, and Rumbaugh have combined their methodologies, together with important concepts from some of the other methodologies, in the Unified Modeling Language. The next step for these methodologies is to adapt them to reuse on a larger scale. Jacobson, working with Martin Griss of Hewlett Packard Laboratories and Patrik Jonsson of Rational Software Corporation, presented a tutorial to this effect at OOPSLA 96 (Object-Oriented Programming Systems, Languages, and Applications).

In the announcement of the tutorial, the three noted that "many people naively equate reuse with objects, expecting it to 'automatically' ensure reuse." There is more to reuse than just inheritance between objects. One further goal is to provide means for reuse, not only at the code level, but upstream. The three describe an approach that begins with business process reengineering and extends through a series of models—requirements, analysis, and design—to the code. Reuse should be able to take off from model elements at any of these levels.

The other major goal is to provide reusable elements at a larger size than small objects. They call these elements component systems. Beyond these methodological requisites, however, are more general organizational obligations. "In almost all cases of successful reuse, management support, simple architecture, a dedicated component group, a stable domain, standards, and organizational support were the keys to success," they said.

References

[1]   P.A. Strassmann, "Information: America's Favorite Investment," *Computerworld*, Aug. 5, 1996, p. 64.

[2]   M.E. Porter, *Competitive Advantage: Creating and Sustaining Superior Performance*, The Free Press, a division of Macmillan, Inc., New York, N.Y., 1985, 557 pp.

[3]   C.K. Prahalad and G. Hamel, "The Core Competence of the Corporation," *Harvard Business Rev.*, May-June 1990, pp. 79–91.

[4]   M. Hammer, "Reengineering Work: Don't Automate, Obliterate," *Harvard Business Rev.*, July-Aug. 1990, pp. 104–112

Chapter 37

The People Process

"The capacity to manage human intellect—and to convert it into useful products and services—is fast becoming the critical executive skill of the age."—James Brian Quinn, Philip Anderson, and Sydney Finkelstein [1]

Software development is a people-intensive process. It follows then, to improve the process, some one, or some thing, or some motivation must improve the people and the process through which people work together.

"Better people," snorts the iconoclast. *"People haven't changed since Homer recounted the goings on at Troy."*

On the contrary, something has changed. A highly complex civilization has evolved. We members of the 132nd generation since Troy do manage it somehow. "Somehow" expresses the ad hoc way in which we now do it.

In the last few hundred years scientists have learned, while working for the most part independently or in small groups, to coordinate work through refereed publications. Medical doctors initiate new doctors into the realities of actual practice through internships and residencies. PhD programs move students from knowledge absorption in classes to knowledge application in dissertations.

Software development differs from much previous work based on knowledge in that scores or hundreds of people often have to work on a single project. The people who can bring extensive knowledge to bear on a complex set of requirements in a large organizational setting are special to begin with:

- They must have a basic education in software.
- They must have the motivation to continue learning, for the projects keep changing and the methods of software development keep advancing.

- Although educated and often experienced as individuals, they must learn to work in teams, and the teams must learn to work with other teams.

- Finally, in spite of being one on a team, and one team among many, at least some of them must motivate themselves to be creative, to break out of the bureaucratic patterns that have characterized organization.

Enhance people's capability

The mindless worker who performed one of Adam Smith's pin-making operations 200 years ago has turned into today's "couch potato." He or she is not a good candidate for a knowledge organization. Just who is a good candidate? Obviously, they should have a level of expertise appropriate to their education and experience.

Bill Gates has called for three additional qualities: ambition, intelligence, and business judgment. Microsoft finds it difficult to meet this standard. "Of the developers interviewed at their universities, Microsoft typically asks only 10 to 15 percent back for additional interviews [in Redmond, Washington], and then hires only 10 to 15 percent of the final group," Michael Cusumano and Richard Selby report [2].

Training. So, you start with the best people you can find, but you have to maintain that specialness. Software technology improves continuously and parts of it change dramatically every few years. People have to learn day-in and day-out to keep up with the constant improvements. Recurrently, software people have to undergo the mental wrench of drastically new methods—third-generation languages replaced assembly language, workstations replaced occasional access to a mainframe, software tools replaced paper and pencil.

In the older engineering fields, executive management expects a bridge to stand or a chemical plant to operate. Yet it does not feel that it has personally to calculate beam stresses or write chemical equations. It has merely to be confident that people educated in such technical processes are executing them according to established methods.

In contrast, software engineering is only a few decades old. Its underlying supports are as yet incomplete, compared to older branches of engineering [3]. Nevertheless, already there is a great deal of knowledge to draw on.

It is largely people who bring this knowledge into the organization. It is training and further education that keeps this knowledge current after hire. However, organizations sometimes carry out training ineffectively—a possibility managers concerned with process improvement might look into [4].

Education concerns itself with what the organization needs to learn to get ready for next year or the year after. It takes place in short courses, graduate study, professional conferences, and workshops.

The only point at issue, ultimately, is whether education and training are working. Are the programs adequate? Are people making use of them?

Mentoring. Education and training bring people only to a level of adequacy. The gap between the knowledge acquired in this way and the ability to apply that knowledge to real-life problems is notorious. Many older professions attempt to bridge that gap through supervised practice. Junior accountants, for example, endure two years of fieldwork in order to become certified public accountants.

Microsoft employs a version of this approach. It has not invested heavily in formal training. It has put new people directly into small work teams. "Team leads, experts in certain areas, and formally appointed mentors take on the burden of teaching in addition to doing their own work," Cusumano and Selby state [2].

According to Quinn and his co-authors, "Professional know-how is developed most rapidly through repeated exposure to the complexity of real problems" [1].

On the one hand, the forced intensity of this approach seems to enhance the confidence of those who complete internships, as the celebrated self-confidence of medical practitioners attests. The software developer, too, soon feels able to attack problems on his own.

On the other hand, the wear and tear of the long hours that typically accompany forced development sometimes leads to burnout, as the media occasionally report when some intern on a 36-hour shift misdiagnoses a patient. That is "quality," or lack of quality. Similarly, our data indicate that software developers make more errors when working under too great time pressure. The methods outlined earlier in this book should enable software organizations to find the happy middle.

Ongoing. In time these special people survive their education, training, and mentoring and become themselves full professionals. That is not the end of the trail. In some of the established professions, the technology that underlies their practice changes rather slowly. That has not been the case with software technology. This rapidly changing field faces the additional pressure of moving entire new technologies into its organizations. One view of this technology transition is that of "technology push," where the originators of the technology try to move it into organizations they feel can profitably use it [5].

Another view might be called "technology pull," where the organization itself pulls in new technology that promises to improve its process. It is the organization's established professionals, of course, who have to play a major role in this "pull." They are the people who have to see in some new technologies the means to a better organization—and that other highly hyped approaches may turn into dead ends.

As "technology pull" takes place, executives with fiduciary responsibilities need to understand the financial implications of what is going on. That means they need to receive measurements along the way, such as an increasing process productivity index. They need to feel that the organization is on the right track so that they can persist in it.

One on one

"Poor management can decrease software productivity more rapidly than any other factor," Barry W. Boehm wrote [6]. At one time we would have put better tools and methods at the head of our list of the factors influencing process productivity. However, our consulting experience leads us to believe that management deserves that honor.

If an organization has good management, it can attain better requirements, quality, tools, and methods. It can select good people, and train and motivate them. It can pull in better technology. With poor management an organization degenerates into the sad experiences with which we have regaled you from time to time.

A puzzling problem is what to do about reaching executives who are above the software technology level. Just what an executive needs to know to oversee his or her particular mix of functions is seldom available in one brief course. He certainly doesn't have time to take courses in each functional area.

Our suggestion is to go one on one. Some of the organization's technologists have the knowledge. An executive or manager can ask a technologist to think through what he ought to know in order to carry out his executive duties with respect to the software field. Have the technologist come to the executive's office for a one-hour one-on-one session.

Don't expect the technologist to have calibrated very precisely what the executive needs to know. He may get into too much detail. He may use unfamiliar words or concepts. The point of being one on one is that the executive can interrupt him. The executive can tell him when he has ventured too deep for his needs. Or the technologist can point out why he thinks the executive needs to know this particular material at this depth. The executive can ask him to explain the unfamiliar. The back-and-forth parley permits both parties to stay on the track better than they could in a large session.

One-on-one encounters are expensive, but about half the cost is the executive's own time. On the one hand, executives don't have very much time to acquire software knowledge, considering the spread of their responsibilities. On the other hand, if some of the decisions they make are flawed for lack of insight into the software world, that cost is high, too.

Attending classes tailored for technologists may appear to be inexpensive, but it may take four or five times as much of the executive's time to sort out what he needs. Much of this class time may be spent on details he doesn't need to know. Of course, there are some subjects where an executive can't get what he needs in one hour of one-on-one parley. When he decides that is the case, he can schedule more sessions.

Also there may be subjects where intensive instruction of more than a few hours is in order. In the case of total quality management, for example, Xerox's chief executive officer decided that each executive level, including himself, was to have a week of full-time training appropriate to its level. Xerox then went to some trouble to sort out what should be included in that week.

Executive briefing

Another medium is the *Executive Briefings* the IEEE Computer Society began publishing in 1996. For 50 years the Society had been publishing conference proceedings, compilations of tutorial papers, journals, and magazines—all addressed primarily to the technology level. It came to realize that many organizations were not keeping up with technology advancements. One of the reasons was that executives at the resource-allocation level no longer understood the needs well enough to provide the capital and training and to inject the necessary impetus.

Our own contribution to this effort is a 25,000 word book called *Executive Briefing: Controlling Software Development*. It focuses particularly on use of the five basic metrics for this purpose [7].

References

[1]   J.B. Quinn, P. Anderson, and S. Finkelstein, "Managing Professional Intellect: Making the Most of the Best," *Harvard Business Rev.*, Mar.-Apr. 1996, pp. 71–80.

[2]   M.A. Cusumano and R.W. Selby, *Microsoft Secrets*, The Free Press, 1995, 512 pp.

[3]   F.P. Brooks, Jr., "Report of the Defense Science Board Task Force on Military Software," Office of the Secretary of Defense, Washington, D.C., 1987, 78 pp.

[4]   M. Shaw, "Prospects for an Engineering Discipline of Software," *IEEE Software*, Nov. 1990, pp. 15–24.

[5] R. Barton, "Technology Transfer Is More Than Training," *American Programmer*, Mar. 1992, pp. 32–37.

[6] B.W. Boehm, "Improving Software Productivity," *Computer*, Sept. 1987, pp. 43–57.

[7] L.H. Putnam and W. Myers, *Executive Briefing: Controlling Software Development*, IEEE Computer Society Press, Los Alamitos, Calif., 1996, 90 pp.

Chapter 38

Leveraging People's Capabilities

"You can tell where your organization is by studying the quality of the thinking, and you can imagine where you want to go by imagining what thinking will be like. Remember: When the thinking changes, the organization changes, and vice versa." —Gerald M. Weinberg [1]

A Roman engineer had to design those wonderful roads, bridges, aqueducts, and buildings with Roman numerals as his tool. Early Mediterranean sea-faring traders had to keep track of their financial state with simple lists; double-entry bookkeeping was not invented until the 14th century in Venice.

Computer hardware price-performance has been improving at a geometric clip for several generations. The computer itself has taken over more of the routine side of programming. But the problems to be programmed continue to grow. Fortunately, software people have software itself to embody the tools that help them.

Capture knowledge in software

"Any encyclopedia or handbook on computing techniques is already beyond what one person could master in a lifetime," Frederick Hayes-Roth, an expert-systems researcher, said [2].

As a result, software-development organizations face these two challenges:

- Keeping what software developers have to do personally within the limits of human capability;
- Improving the process of transferring to developers the technology that still needs to be in their minds.

The solution lies in taking advantage of rapidly evolving computer and software technology.

Our own software life-cycle modeling tools are a case in point. Arithmetic and algebraic operations, statistical and probability computations, curve drawing and visual representation—all have been embodied in the tools. The user does not have to learn these time-consuming disciplines or spend time carrying them out. Essentially all the user needs to do is exercise judgment and evaluate the intangibles in the situation.

Organize information. The management of a technical civilization rests upon information—by now vast quantities of it. Enormous quantities of random information are not of much use. It has to be organized to be useful. Compendiums such as the *Handbook of Chemistry and Physics* make information accessible to knowledgeable users. The established professions have long made use of knowledge, organized in this and other ways.

The capabilities of software extend the ability to organize knowledge. Well-nigh infinite quantities of knowledge can be stored in computer systems, either locally or on a network. This stored knowledge is fundamentally of two types. One is a database—of customers, products, or in our own case, project management data. All parts of an organization—its central headquarters and its scattered plants and sales offices—work off the same up-to-date uniform data.

The second type is tools: spreadsheets, diagram drawing programs, slide preparation programs, as well as tools specialized for software development.

The ability to access and use databases and tools is improving, though there is far to go. Even so, there is more knowledge available than most people have yet learned to use effectively. The organization of this knowledge to facilitate software development is, of course, still in progress.

For the purposes of this chapter, the point is that the ability of software developers to solve problems can be substantially extended by embodying more of the solution methods in software.

Groupware. Except for an occasional brilliant soloist (usually on small projects), software people work in groups. It is also becoming more and more common for them to work with people in widely separated geographical locations. Groupware, such as Lotus Notes or the Coordinator by Action Technologies, facilitates their ability to work together on a common project from nearby offices or different countries [3].

Front-line support. The traditional form of organization in the mass-production factories that grew out of Adam Smith's division-of-labor principle was a topdown structure of executives, managers, and supervisors coordinating the people who did the physical work. With many different jobs, stemming from this division of labor, there was much coordination to do.

Knowledge work is less amenable to coordination by an extensive hierarchical structure. To solve a problem the software developer has to know what the problem is. The customer or the user knows what the problem is, that is, what the requirements are. The developer needs to be in contact with the people who have first-hand experience of the problem. A large supervisory structure tends to filter that knowledge, as it passes from organizational silo to silo or from level to level.

Recognition of these circumstances has led to the "inverted organization." The people who do the work are regarded as the front line that has to be in contact with the user. The former on-top management structure turns into an on-the-bottom support structure. Many of its former functions remain—supplying desks, equipment, building, finances needed to do the work. But its attitude changes. It is now supporting the people, not managing them.

Support makes a difference

If you have not been doing much to improve your software process, you may be mired at what the Software Engineering Institute labels the "chaotic" level. That is called, more politely, the Initial level, or Level 1.

By moving from the primitive methods of the 1960s and 1970s to well established modern methods, you can improve your process productivity index by one or two points, sometimes by three points, as the experience of the Naval Systems division demonstrates.

This division is a unit of a multinational corporation. In 1985 it underbid, or underestimated the difficulty, of a large system to be installed on ships of a NATO country. The system, estimated at 575,000 new SLOC, was to acquire data from all the ship's sensors, process the data, and direct the ship's weapons. The software cost was in the hundred million dollar range.

At the time the division called us in, management was obviously worried. Money was going out at the rate of over $100,000 a day. They felt the project was in trouble. Our analysis of their ongoing data showed they would need some $25 to $30 million more than the bid amount and about two years more schedule. Subsequently the customer agreed to revise the contract at a fixed price of $125 million and a longer schedule.

A little later the corporation sent Ned Philips, one of its best software development managers, to reorient the project. His very first conclusion was that the people did not believe there was any chance they could finish the job on time. Philips spent his first three months largely in listening to them. He feels that there are two dimensions to software management. One is the technology; it is comparatively easy. The other is the people side, the way people feel about the work—their hopes and their fears. Doing something about that is difficult.

In effect, that is what he was getting under way those first three months when all he appeared to be doing was sitting around talking to people. He was helping them see that, yes, there was something wrong, but they had done many things right. He helped them realize that it is much more productive to focus on what is right and improve it, instead of looking at what is wrong and trying to fix it. After all, you always get more of what you focus on.

Philips did one other thing right away. The specifications called for defects remaining at the time of delivery of 0.014 defects/KSLOC. That was wildly unrealistic. He called around to friends doing software development-maintenance work on similar systems and found they were working toward 3 or 4 defects/KSLOC. Apparently the customer's engineers had set the 0.014 figure by analogy to hardware levels, not realizing that software was in another ballpark.

Using common sense, Philips persuaded the customer to change the delivery specification. The people on the project realized the development now could be completed. They stopped fretting and set to work with renewed enthusiasm.

Ned did some other concrete things. For instance, he brought in better testing tools to find defects in development practices. He saw to it not only that developers fixed the defects, but management fixed the practices, too. He tracked the current management numbers and told the people where they stood. He brought in outside assistance to help the people realize that they knew what they needed and how to do it.

This project is the only instance of which we are aware in which the change in direction had so profound an effect. In the early part of the project the productivity index was 8 or 9, in the average range of this type of project. After the mid-stream overhaul the turnaround boosted the productivity index by four points. The reason was Ned Philips' leadership skills. Leadership makes a difference. The numbers at the end of the project were:

2.5 calendar months early
$5 million under the fixed price
0.56 defects remaining/KSLOC
and, by the way, they delivered almost 1,000,000 new SLOC.

Ed Tilford, another experienced project manager put it this way: "[Decision-makers] need to understand the major variables that drive a software project to success. They need to be able to turn these variables into an effective project plan. A sound plan provides the foundation on which people can build—and feel good about themselves and their work" [4].

Keeping up

That brings us to the thought that this book was written at a point in time. Any references we can cite antedate that point. But software development continues to advance. You will need to seek information long after this book is gathering dust on your shelf.

How do you keep up? "The statistics about reading are particularly discouraging," Tom DeMarco and Timothy Lister observed. "The average software developer, for example, doesn't own a single book on the subject of his or her work, and hasn't ever read one" [5].

Oh, my! But you have read this far in this book, so you are well above average. It looks like you are going to have to provide some kind of in-house training for all those other guys on whatever it turns out they need to know.

There are basically four ways of keeping up: periodicals, books, meetings, and contacts.

> **Periodicals.** There are trade magazines, newsletters, society magazines, and transactions and journals, generally in ascending order of reading difficulty. *The American Programmer* is an example of a newsletter. *IEEE Software, Computer*, and *Communications of the ACM* are society magazines that consider software matters. *Transactions on Software Engineering* publishes research papers.
>
> Subscriptions to periodicals, often involving membership in one of the professional societies, have the incidental merit of getting you on mailing lists that tell you about new books and meetings.
>
> **Books.** Books exist on several levels. First are textbooks, intended for use in college and extension classes. Often they are a good place to start when you are not familiar with a subject. Second are books addressed to working practitioners, such as the present one.
>
> Third are so called tutorial books—collections of previously published articles and papers on the same subject—prepared by the IEEE Computer Society, the Association for Computing Machinery, and occasionally commercial publishers. They are not always a good place to start, as the papers may be rather advanced. Sometimes they start off with simple introductory articles that are helpful. They are a convenient and inexpensive way of get-

ting access to the significant articles and papers on a given topic—especially if you don't live next to a technical library.

Meetings. There are an infinite variety of meetings: workshops, short courses, user gatherings, society conferences, industry conferences, and trade shows. One conference that deals with the subject matter of this Part is the International Conference on Software Engineering, held in the United States one year and in Europe or East Asia the next year.

Your organization could profit from a representative at one or two meetings a year on each subject you are in the process of doing something about. A less expensive way of getting part of the advantage of attending a meeting is to buy its proceedings. Another cost-saving approach is to bring an expert or two to your facility, instead of sending your people to the meetings where they appear.

Contacts. The other advantage of meetings is that you meet people at them. You get "the real skinny" of what live people are actually doing. Sometimes the written word gets overly enthusiastic. People shrink from publishing their problems.

One company's experience

One of our clients, which we shall call Exceptional Programming Services, has achieved productivity indexes as high as 30 and 32 on business systems. Business systems are now being developed by those reporting to us at an average productivity index of 16.9. Exceptional Programming had been attaining a PI of 18 on systems programmed in Cobol, using a database management system. Then it invested $15 million and five years in developing in-house an integrated computer-aided software engineering tool. Most of the input to this tool is diagrams and fill-in screens.

When first used on the development of a real payroll system, this tool hit a PI of 30. With more experience Exceptional Programming reached PIs of 32 on later projects. A PI of 32 is 39 times as productive (in terms of process productivity) as a PI of 16.9 (the business systems average in 1995).

This tool, still a prototype, has some drawbacks. For example, the amount of code it generates is about 30 percent greater than the hand-coded amount would be. This increase in size is no great problem in the case of the large mainframes with tremendous amounts of virtual memory on which these programs run. Even so, about 20 percent of the tool-generated Cobol (mostly computational algorithms) has to be either hand-written or reworked.

Exceptional Programming's experience shows that huge gains in productivity can be attained by organizations able to incorporate the latest technology in their tools. Note that it does take years of time and millions of investment dollars. There is "no silver bullet," as Brooks expressed it. There is rapid payback.

The plodding organizations are not going to assimilate tools of this caliber easily because they are hard to use. The learning curve is steep. These advanced tools apply only to simple, routine systems at present. They may become even more efficient when we turn the complicated, but routine, functions found in these applications into reusable component systems.

References

[1] G.M. Weinberg, *Quality Software Management, Vol. 2, First-Order Measurement*, Dorset House Publishing, New York, N.Y., 1993, 346 pp.

[2] B. Chandrasekaran, "Interviews: Frederick Hayes-roth and Richard Fikes," *IEEE Expert*, Oct. 1991, pp. 7–14.

[3] T. Winograd, "Groupware: Systems Design from Perspective of Getting Things Done," *IEEE Software*, Nov. 1991, pp. 81–82.

[4] E. Tilford, "Successful Projects Are Built On People, Planning, and Flow," *Computer*, Mar. 1992, pp. 94—95.

[5] T. DeMarco and T. Lister, *Peopleware: Productive Projects and Teams*, Dorset House Publishing Co., New York, N.Y., 1987, 188 pp.

Chapter 39

Get Up-to-Date

"Few fields have so large a gap between current best practices and average current practice." —Frederick P. Brooks [1]

If you have not been doing much to improve your software process, you may be mired at what the Software Engineering Institute labels the "chaotic" level. That is called, more politely, the Initial level, or Level 1. If you do not feel ready to undertake the steps outlined in the last few chapters, you might consider moving from the primitive methods of the 1960s and 1970s to the well-established methods outlined below. By adopting methods such as these, you can improve your process productivity index by one or two points, sometimes by three points.

Use modern programming practices

A now classic conference introduced the term software engineering in 1968—quite a while ago. The idea was that the orderly, disciplined ways of working that older branches of engineering had developed also would be effective in improving the software development process. Some of these methods, summarized in Table 39-1, came to be known collectively as modern programming practices.

Various studies in the 1970s asserted that these practices reduced both cost of development and defects by from 25 to 75 percent [2]. Use of these methods did not spread as rapidly as the potential gain in productivity would suggest. The methods were new, of course, and complex. They had to be disseminated by "soft" methods, such as education and training. That took time, cost money, and required organizational discipline.

By now, tool-builders have incorporated modern programming practices in commercial tools. You may already be practicing some of them. Still, there may be some from which you could still benefit.

Table 39-1. Some of the methods included within the general term modern programming practices are:

Structured design	Partitions large systems into independent subsystems that relate to each other in a structured way
Modular decomposition	Hides details of operation within a module. Relates modules to each other with carefully defined control terms and data transfers. Except for the designer of the module interior, other project members need understand only the external transfers
Top-down development	Partitions and decomposes the proposed system into a hierarchy of subsystems and modules, beginning at the top and working down
Structured programming	Structures the flow of logic from one instruction to the next without arbitrary branching, making it easier for later programmers to understand
Structured walkthroughs	Reviews programs developed by structured methods with emphasis on defect detection, not blame. Management does not attend. The developer being reviewed is responsible for corrections

Use third-generation languages

If you are still doing some of your programming in second-generation assembly languages, move up to third-generation languages, such as Cobol, Fortran, Pascal, C, C+, C++, and Ada. One instruction in a third-generation language replaces three to five instructions in a second-generation language. The improvement in coding productivity is in about the same ratio.

For instance, Capers Jones estimated the same program in several languages. It required 20,000 SLOC in assembly or 6,500 SLOC in Cobol, a ratio of 3.08 to 1. The cost-of-coding ratio was 3.25 to 1. The overall cost ratio from requirements through field maintenance was 2.31 to 1 [3].

Years ago assembly language was all there was. Then, for several decades after the arrival of the first third-generation languages, assembly language could still run critical algorithms faster. Also programs written in assembly language were shorter. Hence they could fit in less memory and in those days memory was often at a premium. Now, with the high density of semiconductor chips, processing power and memory capacity are abundant and inexpensive. There is seldom need any more for these capabilities of assembly language. Priority has shifted to reducing programming costs.

Use fourth-generation languages

Spreadsheet programs are an example of a software package that enables its nonprogrammer users to generate what is, in effect, a program. By entering numbers and formulas into the spreadsheet format, users themselves create a program to manipulate their banks of numbers. Fourth-generation languages operate in a comparable fashion. Users answer questions on input screens, for example, and the language generates code.

These packages originated in the business field and may be better known by their commercial names, such as Nomad, Ramis, Oracle, Mantis, Informix, Micro Focus, ADF, and dBase. They generally include components for managing a database, providing screens, generating reports, and preparing graphic outputs. Typical users experience a three- or fourfold improvement in productivity over a third-generation language.

Where a third-generation language requires a programmer to detail how the application is to be performed, a fourth-generation language permits a nonprogrammer merely to indicate what is to be done.

To gain their advantages, fourth-generation languages have to be specialized to particular tasks. "A user can save application-development time if the problem matches the assumptions in the tool's predefined nonprocedural facilities," Santosh K. Misra and Paul J. Jalics reported, following the comparative development of a sample problem in two fourth-generation languages and Cobol. "If the problem is not the kind the tool was designed for, the user may pay development and performance penalties" [4].

Robert L. Glass's review of six studies led him to conclude that "the benefits are evolutionary rather than the oft-claimed revolutionary" [5].

Use application generators

"Application generators translate specifications into application programs," reported J. Craig Cleaveland [6]. Application generators are related to fourth-generation languages because the specification is sometimes written in a fourth-generation language.

They increase developer productivity and reduce defects. But they are usually narrowly applicable. Thus, potential users must make a balance between the advantages and the cost.

Consider outsourcing

One step beyond buying fourth-generation languages or application generators is outsourcing, or contracting with an outside organization to provide an entire data processing service. The vendor provides both hardware and pertinent software for your application. In effect, you are reusing some of the vendor's existing software [7].

Move to interactive development

Again, in the beginning the mainframe was all there was. Every programmer had to take turns compiling and running his newly written code on the organization's one mainframe. He probably had to wait several days, not only losing time, but also losing his train of thought. In the 1960s programmers hailed time sharing as a step forward. In the 1970s the minicomputer made processing power more accessible. In the 1980s the personal computer arrived, followed by the workstation.

Wherever you are on this line of advance, move up to the appropriate workstation as soon as you can. You can not only compile and try out your programs whenever they are ready, but also utilize software tools on your workstation that makes your work more efficient [8].

Computer-aided software engineering

One step beyond individual (and perhaps incompatible) commercial tools is CASE (computer-aided software engineering). Variations of the concept include Integrated CASE, integrated project-support environment (IPSE), and integrated software-engineering environment (ISEE). The idea of CASE dates back to the mid-1980s, but commercial implementations are more recent [9]. All are still far from the ultimate system.

The general idea is to automate much of the detail work in software development, both to alleviate the shortage of qualified people and to improve product quality. By making tools work together, integration makes them easier to use [10]. Even so, many organizations have had trouble getting their people to use CASE implementations.

Nevertheless, versions of it are here [11] [12]. Forward-looking organizations are using them successfully.

Users have learned three lessons:

- Try out a promising CASE tool with a few experienced people.
- When they seem satisfied, extend its use to an entire project and perhaps later to an entire organization.

- Expect to expend substantial sums—on the order of the cost of the tools themselves—on indoctrination and training [13].

Document efficiently

The cost of the words written on a project is usually greater than the cost of the instructions coded [3]. We all like to decry "paper" and "red tape." If the customer, user, or nature of the product itself requires documentation—specifications, detailed design, test plans, change orders, manuals, and trouble reports—stop groaning. They are an established part of the product to be produced and they deserve to be produced efficiently.

Get the personal computers or workstations, word processors, databases, spreadsheets, and more specialized software packages (such as CASE) that enable your people to produce, store, access, and maintain paperwork.

Client-server systems

"Large, central computer systems with proprietary architectures [that is, the long famed mainframe] are being replaced by distributed networks of low-cost computers in an open systems environment [that is, personal computers and workstations]," says Pieter R. Mimno, writing in a theme issue of *American Programmer* [14]. A client-server system consists of a collection of personal computers or workstations on a local or wide area network plus a server connected to the common database. With the rapid development of microprocessors, such a system can bring more computing power to bear than mainframes of a few years ago.

This change is having an effect on software. At the personal computer level, the graphical user interface (such as Windows) and the operating system are commercially available. Application programs number in the tens of thousands. Users need seldom concern themselves with new programming. Similarly, at the client-server level, much software is commercially available. Users still need to program functions unique to their activities.

Many tools make this new programming more efficient. "Most commercial software organizations need development tools to help build network applications with graphical user interfaces (GUIs), relational databases access, and an event-driven use model," Paul Bloom said [15]. To the extent that users can conform their activities to commercial programs, they can avoid developing their own programs.

References

[1]   F.P. Brooks, Jr., "Report of the Defense Science Board Task Force on Military Software," Office of the Secretary of Defense, Washington, D.C., 1987, 78 pp.

[2]   W. Myers, "The Need for Software Engineering," *Computer*, Feb. 1978, pp. 12-26.

[3]   C. Jones, *Programming Productivity*, McGraw-Hill Book Co., New York, N.Y., 1986, 280 pp.

[4]   S.K. Misra and P.J. Jalics, "Third-Generation versus Fourth-Generation Software Development," *IEEE Software*, July 1988, pp. 8–14.

[5]   R.L. Glass, "A New Look At The Numbers: CASE and 4GLs: What's The Payoff," *The Software Practitioner,* Jan. 1991, pp. 5–8.

[6]   J.C. Cleaveland, "Building Application Generators," *IEEE Software*, July 1988, pp. 25–33.

[7]   M.C. Lacity and R. Hirschheim, "The Information Systems Outsourcing Bandwagon," *Sloan Management Rev.,* Fall 1993, pp. 73–86.

[8]   R. Comerford, "Focus Report and Guide To Engineering Workstations and PCs," *IEEE Spectrum*, May 1993, pp. 35–77.

[9]   E.J. Chikofsky, "Software Technology People Can Really Use," introducing theme issue on CASE, *IEEE Software*, Mar. 1988, pp. 8–10.

[10] R.J. Norman and M. Chen, "Working Together to Integrate CASE," introducing theme issue on CASE, *IEEE Software*, Mar. 1992, pp. 12–16.

[11] E.J. Chikofsky, D.E. Martin, and H. Chang, "Assessing the State of Tools Assessment," introducing theme issue on tools, *IEEE Software*, May 1992, pp. 18–21.

[12] A. Topper, "Automating Software Development," *IEEE Spectrum*, Nov. 1991, pp. 56–62.

[13] C. Jones, "CASE's Missing Elements," *IEEE Spectrum*, June 1992, pp. 38–41.

[14] P.R. Mimno, "Client-Server Computing," *American Programmer*, Apr. 1993, pp. 19–26.

[15] Bloom, "Trends in Client Server/Cooperative Processing Application Development Tools," *American Programmer*, Apr. 1993, pp. 27–33.

Chapter 40

You Make it Happen

"Knowing is not enough; we must apply. Willing is not enough; we must do." —Goethe

"These ideas are just too confusing," John complained. "Lengthen this and that goes down. Improve something else and everything else goes down. Who do they think we are? Einstein?"

"I told the two author guys that you were getting mixed up," I said. "I suggested that these ideas would go down better if they would let me, Honest Phil, put my special twist on them. Credibility is very important, I said."

"You, Honest Phil?" John exclaimed. "I've been called Honest John since kindergarten, but everybody calls you Foxy Phil."

"Wily maybe, but not sly," I riposted. "You'll admit that the length of schedule has a lot to do with what happens on a project?"

"I know that if the people are good, they'll get the work done faster," John replied.

"We'll get to that. Let's deal with development time first," I said, pulling up a piece of paper. "Now, if we plan a schedule shorter than the minimum development time, we land in the Impossible Region." I drew this possibility on the paper (Table 40-1).

"I've been there," John said. "I want to stay out of it."

"Second, if you plan a schedule at the minimum development time, you can do it, but effort, cost, team size, and number of defects are all at the maximum. MTTD, as the reciprocal of defects per month, is at a minimum."

Table 40-I. There are only five actions you can employ in planning a software schedule.

Action	Effect				
Plan Schedule	Development Time	Effort Cost	Team Size	Number of Defects	MTTD
Below Minimum	Impossible Region				
At Minimum	Min	Max	Max	Max	Min
Longer than minimum	⇑	⇓	⇓	⇓	⇑
Beyond 130 percent of Minimum	Impractical Region				
Extend Schedule Beyond FOC*	⇑	⇑	—	⇓	⇑

*Full Operational Capability

"Yeah, minimum development time is not a very good place to be," John said. "I've been there, too."

"Well, it's the knife-edge limit," I said, "and a knife edge is not a comfortable place to be."

I sketched in the third line. "The range of schedules longer than the minimum is really the operating region. As you plan a longer schedule, within reason, the other management numbers go down—except MTTD, which goes up."

"As a working stiff, this is where I want to be," John declared.

"You can plan a still longer schedule, beyond some 130 percent of the minimum, but that schedule is likely to land you in the Impractical Region. Out there, there are too few people on the project to cover resignations or unexpected problems. The time may be unreasonably long in terms of getting to market fastest."

"I was on a project that they canceled because several other companies got their products out there first," John offered.

"Finally, one might extend the schedule beyond the full-operational-capability milestone," I said, penciling in the last line. "Coding is complete, but there are still a lot of defects lurking in it. Extending the schedule raises costs, but gives the development team time to root out the remaining defects. They are better at that than a new team of maintainers would be. That raises MTTD, of course."

"Hunting for defects is kind of fun," John observed. "I often wish we had more time budgeted for doing that."

"Looking for causes is even more fun than looking for defects," I rebutted.

"Fixing process flaws would be still more fun," John asserted, topping me.

"Hush, this is a serious book. You're not supposed to talk about fun."

"All right, it's no fun to build up a project so fast everybody is stumbling all over each other," John came back.

I took out another piece of paper. "Another action that operates on the project time scale is manpower buildup. If you build up faster, development time is naturally shorter, but effort, cost, team size, and number of defects all go up. MTTD is worse (Table 40-2)."

"That's not good," John noted. "Why do managers try to speed up the buildup then?"

"Because they live in a competitive world, dummy. They are under pressure to get the product out so the company can start making money with it."

"They could save money by building up more slowly," John said as I filled in the second line of the chart.

"That's a tradeoff they have to make," I replied. "Often the savings are small compared to the gain from using the product. However, managers do not have complete freedom to set the manpower buildup rate at the particular level that best suits their tradeoff needs. To a considerable extent the manpower buildup rate is going to be the same on the next project as it was on the last project."

Table 40-2. There are only two approaches, faster and slower, to building up manpower.

Action	Effect				
Manpower Buildup	Development Time	Effort Cost	Team Size	Number of Defects	MTTD
Build Up Faster	⇓	⇑	⇑	⇑	⇓
Build Up Slower	⇑	⇓	⇓	⇓	⇑

"I remember that you found the previous rate by calibration which, I believe, you learned in college," John scoffed. "It seems to me that a manager should be able to build up at whatever rate he has people to support."

"She can throw people at a project as fast as she likes," I replied. "The point is they can be usefully employed only at the rate that little chunks of work can be separated out for each new person. The rest just get in the way. Setting up the little chunks depends on the nature of the work, whether it is mostly sequential or mostly concurrent, things like that. Usually the nature of the work does not change much from one project to the next in the same organization."

"Still, he could add people more slowly than the chunks become available," John objected. "Then he could get on the bottom line of your table and save all those good things."

"That's true so far as the project itself is concerned. But she is still up against her particular set of customers or users. They are accustomed to getting a new product in a certain time frame. The manager can't arbitrarily push them into a longer time frame. She has to explain the cost advantages of moving a little slower and perhaps sell them on a slower buildup."

"There you go again," John said accusingly. "Every time it makes sense to me to follow one of the lines in your table, you wiggle off the hook. That's why they call you Foxy Phil."

"The table is simple enough," I pointed out. "This one only has two lines. It's all the human beings we have to deal with that make software development complicated."

I pulled out another sheet of paper. "Sometimes the customer needs a product in a time that turns out to be less than the minimum development time. You just can't do the entire product he wants on the schedule he wants."

"Tell him to go to hell," John suggested.

"Temper, temper. First, we explain why we have a minimum development time. Then we ask him if he could get by in the early days of the new system with reduced functionality—skip the bells and whistles in the first release. If you can reduce the size of the proposed system, then all the management numbers go down, like this (Table 40-3). Conversely, if you increase size, they all go up."

I could see that John was getting restless. "This is the last table," I said, taking out another sheet. "If an organization can improve its process productivity, all the management numbers get better." I filled

Table 40-3. If you have limited time or money, you can reduce the product's initial functionality.

Action	Effect				
Functionality	Development Time	Effort Cost	Team Size	Number of Defects	MTTD
Reduce Size	⇓	⇓	⇓	⇓	⇑
Increase Size	⇑	⇑	⇑	⇑	⇓

in the first line (Table 40-4). "Obviously, this is the smart thing to do. The only drawback is that it takes time, usually time measured in years."

"There is one other little problem," John said. "Improving process productivity is not easy. You've got to think of everything under the sun."

"That's why they put the word 'total' in 'total quality management.' You try to improve everything about the process, including the people and the technology."

"And not everybody succeeds," John said, pointing to the second line of the table.

"I just put that line in for the sake of completeness. Organizations don't slide back very often. Sometimes they do get a new division general manager who cuts process investment, insists upon fast deliveries, and scares away the best people. Then all the numbers get worse."

"The human race discourages me at times," John said, wiping away a crocodile tear. "I notice that you foxily put each of those situations on a separate sheet of paper so I would think it was very simple. In the

Table 40-4. Improving process productivity is the best course. In this way all the management numbers get better.

Action	Effect				
Process Productivity	Development Time	Effort Cost	Team Size	Number of Defects	MTTD
Improve	⇓	⇓	⇓	⇓	⇑
Decline	⇑	⇑	⇑	⇑	⇓

real world all those factors—11 of them—are operating simultaneously, and each is having an up or down effect on the management numbers."

"You caught me," I admitted, "but 11 is better than 22 or 44. You can count up an awful lot of factors in software development. Even with 11, you need some kind of analytical framework to show their relationships."

"You didn't say anything about the fact that the management numbers, at the planning stage, are only "expected" numbers. The probability of achieving an expected projection is only 50 percent."

"Hey, you understand this stuff better than you let on, and you call yourself Honest John. You've just been leading me on," I blurted.

"Well, Foxy, we all have trouble keeping these relationships straight in our heads. The way you explain it, we can make it happen."

On a foundation of sound measurement, we will carry out well-planned projects, building excellent products, with a continuously improving process.

"The problems of software management don't touch me anymore since I took up world empire."

Bibliography

J. August, *Joint Application Design*, Yourdon Press, Prentice Hall, Inc., Upper Saddle River, N.J., 1991.

N.R. Augustine, *Augustine's Laws*, Penguin Books, New York, 1987, 484 pp.

P.B. Bassett, "To Make or Buy? There is a Third Alternative," *American Programmer*, Nov. 1995.

P.G. Bassett, *Frame-Based Software Engineering*, Prentice Hall Inc., Upper Saddle River, N.J., 1996, 384 pp.

P. Bloom, "Trends in Client-Server/Cooperative Processing Application Development Tools," *American Programmer*, Apr. 1993, pp. 27–33.

B.W. Boehm, *Software Engineering Economics*, Prentice-Hall Inc., Englewood Cliffs, N.J., 1981, 767 pp.

B.W. Boehm, "Improving Software Productivity," *Computer,* Sept. 1987, pp. 43–57.

B.W. Boehm, "A Spiral Model of Software Development and Enhancement," *Computer*, May 1988, pp. 61–72.

B.W. Boehm, *Tutorial: Software Risk Management*, IEEE Computer Society Press, Los Alamitos, Calif., 1989, 496 pp.

G. Booch, *Software Engineering With Ada,* Benjamin-Cummings Publishing Co., Menlo Park, Calif., 1983, 504 pp

F.P. Brooks Jr., *The Mythical Man-Month: Essays on Software Engineering,* Addison-Wesley Publishing Co., Reading, Mass., 1974, 195 pp.

F.P. Brooks, Jr., "No Silver Bullet: Essence and Accidents of Software Engineering," *Computer*, Apr. 1987, pp. 10–19.

F.J. Buckley, *Implementing Configuration Management: Hardware, Software, and Firmware,* IEEE Press, Piscataway, N.J., 1992, 256 pp.

E.J. Chikofsky, "Software Technology People Can Really Use," introducing theme issue on CASE, *IEEE Software*, Mar. 1988, pp. 8–10.

E.J. Chikofsky, D.E. Martin, and H. Chang, "Assessing the State of Tools Assessment," introducing theme issue on tools, *IEEE Software*, May 1992, pp. 18–21.

R.H. Cobb and H.D. Mills, "Engineering Software under Statistical Quality Control," *IEEE Software*, Nov. 1990, pp. 44–54.

S.D. Conte, H.E. Dunsmore, and V.Y. Shen, *Software Engineering Metrics and Models*, The Benjamin/Cummings Publishing Co., Inc., Menlo Park, Calif., 1986, 396 pp.

M.A. Cusumano, *Japan's Software Factories: A Challenge to U.S. Management*, Oxford University Press, New York, N.Y., 1991, 513 pp.

M.A. Cusumano and R.W. Selby, *Microsoft Secrets*, The Free Press, 1995, 512 pp.

T. DeMarco, *Controlling Software Projects*, Yourdon Inc., New York, N.Y., 1982. 284 pp.

T. DeMarco and T. Lister, *Peopleware: Productive Projects and Teams*, Dorset House Publishing Co., New York, N.Y., 1987, 188 pp.

R. Dion, "Elements of a ·Process-Improvement Program," *IEEE Software*, July 1992, pp. 83–85.

R. Dion, "Process Improvement and the Corporate Balance Sheet," *IEEE Software*, July 1993, pp. 28–35.

M. Dorfman and R.H. Thayer, *Standards, Guidelines, and Examples on System and Software Requirements Engineering*, IEEE Computer Society Press, Los Alamitos, Calif., 1990, 620 pp.

J.B. Dreger, *Function Point Analysis*, Prentice-Hall, Englewood Cliffs, N.J., 1989.

P.F. Drucker, *The Effective Executive,* Harper & Row, New York, N.Y., 1966, 178 pp.

P.F. Drucker, *Management: Tasks, Responsibilities, Practices*, Harper & Row, New York, N.Y., 1973, 839 pp.

M.E. Fagan, "Design and Code Inspections to Reduce Errors in Program Development," *IBM Systems J.,* Vol. 12, No. 3, 1976, pp. 219–248.

D.C. Gause and G.M. Weinberg, *Exploring Requirements: Quality Before Design*, Dorset House Publishing, New York, N.Y., 1989, 320 pp.

S.L. Gerhart, "Applications of Formal Methods: Developing Virtuoso Software," introducing a theme issue of *IEEE Software*, Sept. 1990, pp. 7–10.

T. Gilb and D. Graham, *Software Inspections*, Addison-Wesley Publishing Co., Reading, Mass., 1993, 471 pp.

R.B. Grady and D. Caswell, *Software Metrics: Establishing a Company-Wide Program*, Prentice Hall, Englewood Cliffs, N.J., 1987, 288 pp.

R.B. Grady, *Practical Software Metrics for Project Management and Process Improvement*, Prentice Hall, Englewood Cliffs, N.J., 1992, 270 pp.

M. Hammer, "Reengineering Work: Don't Automate, Obliterate," *Harvard Business Rev.,* July-Aug. 1990, pp. 104–112.

D.J. Hatley and I.A. Pirbhai, *Strategies for Real-Time System Specification*, Dorset House Publishing, New York, N.Y., 1987, 412 pp.

J. Herbsleb et al., "Software Process Improvement: State of the Payoff," *American Programmer*, Sept. 1994, pp. 22–12.

W.S. Humphrey, *Managing the Software Process*, Addison-Wesley Publishing Co., Reading, Mass., 1989, 494 pp.

W.S. Humphrey, *A Discipline for Software Engineering,* Addison-Wesley, Reading, Mass., 1995, 789 pp.

I. Jacobson, M. Ericsson, and A. Jacobson, *The Object Advantage: Business Process Reengineering with Technology,* Addison-Wesley Publishing Co., Reading, Mass., 1995, 347 pp.

C. Jones, *Programming Productivity*, McGraw-Hill Book Co., New York, N.Y., 1986, 280 pp.

C. Jones, *Assessment and Control of Software Risks*, Prentice Hall Inc., Upper Saddle River, N.J., 1993.

M.M. Lehman and L.A. Belady, *Program Evolution: Processes of Software Change*, Academic Press, New York, N.Y., 1985, 538 pp.

D.H. Longstreet, *Software Maintenance and Computers*, IEEE Computer Society Press, Los Alamitos, Calif., 1990, 294 pp.

R.B. Mays, "Defect Prevention and Total Quality Management," in *Total Quality Management for Software*, edited by G. Gordon Schulmeyer and James I. McManus, Van Nostrand Reinhold, New York, N.Y., 1992, 497 pp.

D. McNeill and P. Freiberger, *Fuzzy Logic*, Simon & Schuster, New York, N.Y., 1993, 319 pp.

P.V. Norden, "Useful Tools for Project Management," from *Operations Research in Research and Development,* edited by B.V. Dean, John Wiley & Sons, New York, N.Y., 1963.

R.J. Norman and M. Chen, "Working Together to Integrate CASE," introducing theme issue on CASE, *IEEE Software*, Mar. 1992, pp. 12–16.

K. Orr, *The One Minute Methodology*, Dorset House Publishing Co., New York, N.Y., 1984, 59 pp.

M.E. Porter, *Competitive Advantage: Creating and Sustaining Superior Performance*, The Free Press, a division of Macmillan, Inc., New York, N.Y., 1985, 557 pp.

C.K. Prahalad and G. Hamel, "The Core Competence of the Corporation," *Harvard Business Rev.*, May-June 1990, pp. 79–91.

L.H. Putnam, "A General Empirical Solution to the Macro Software Sizing and Estimating Problem," *IEEE Trans. Software Eng.,* July 1978, Vol. SE-4, No. 4, pp. 345–361.

L.H. Putnam and W. Myers, *Measures For Excellence: Reliable Software On Time, Within Budget,* Prentice-Hall, Inc., Englewood Cliffs, N.J., 1991, 372 pp.

L.H. Putnam and W. Myers, "Haste Makes Waste: The Way You Plan A Project Affects Its Reliability," *American Programmer*, Feb. 1992, pp. 21–27.

L.H. Putnam and W. Myers, "Penetrating the Metrics Smoke Screen," *American Programmer*, Dec. 1995, pp. 14–19.

L.H. Putnam and W. Myers, *Executive Briefing: Controlling Software Development,* IEEE Computer Society Press, Los Alamitos, Calif., 1996, 90 pp.

J.B. Quinn, P. Anderson, and S. Finkelstein, "Managing Professional Intellect: Making the Most of the Best," *Harvard Business Rev.*, Mar.-Apr. 1996, pp. 71–80.

B.E. Swanson and C.M. Beath, *Maintaining Information Systems in Organizations*, John Wiley and Sons, New York, N.Y., 1989.

R.H. Thayer and M. Dorfman, *System and Software Requirements Engineering*, IEEE Computer Society Press, Los Alamitos, Calif., 1990, 735 pp.

C.E. Walston and C.P. Felix, "A Method of Programming Measurement and Estimation," *IBM Systems J.*, Vol. 16, No. 1, 1977, pp. 54–73.

T.J. Watson Jr. and P. Petre, *Father Son & Co.: My Life at IBM and Beyond*, Bantam Books, New York, N.Y., 1990, 468 pp.

G.M. Weinberg, *Quality Software Management: Volume 1, Systems Thinking*, Dorset House Publishing, New York, N.Y., 1992, 318 pp.

G.M. Weinberg, *Quality Software Management, Volume 2, First-Order Measurement*, Dorset House Publishing, New York, N.Y., 1993, 346 pp.

E.F. Weller, "Lessons from Three Years of Inspection Data," *IEEE Software*, Sept. 1993, pp. 38–45,

E. Yourdon and L.L. Constantine, *Structured Design: Fundamentals of a Discipline of Computer Program and Systems Design*, Prentice-Hall, Englewood Cliffs, N.J., 1975, 473 pp.

E. Yourdon, *Rise & Resurrection of the American Programmer*, Yourdon Press, Prentice Hall, Upper Saddle River, N.J., 1996, 318 pp.

Index

IEEE Computer Society Press Editorial Board

Practices for Computer Science and Engineering

Editor-in-Chief
Mohamed E. Fayad, University of Nevada

Associate Editor-in-Chief
Nayeem Islam, IBM T.J. Watson Research Center

The IEEE Computer Society now publishes a broad range of practical and applied technology books and executive briefings in addition to proceedings, periodicals, and journals. These titles provide practitioners and academicians alike the tools necessary to become "instant experts" on current and emerging topics in computer science.